Studies in Computational Intelligence

Volume 706

Series editor

Janusz Kacprzyk, Polish Academy of Sciences, Warsaw, Poland
e-mail: kacprzyk@ibspan.waw.pl

About this Series

The series "Studies in Computational Intelligence" (SCI) publishes new developments and advances in the various areas of computational intelligence—quickly and with a high quality. The intent is to cover the theory, applications, and design methods of computational intelligence, as embedded in the fields of engineering, computer science, physics and life sciences, as well as the methodologies behind them. The series contains monographs, lecture notes and edited volumes in computational intelligence spanning the areas of neural networks, connectionist systems, genetic algorithms, evolutionary computation, artificial intelligence, cellular automata, self-organizing systems, soft computing, fuzzy systems, and hybrid intelligent systems. Of particular value to both the contributors and the readership are the short publication timeframe and the worldwide distribution, which enable both wide and rapid dissemination of research output.

More information about this series at http://www.springer.com/series/7092

Gabriele Valentini

Achieving Consensus in Robot Swarms

Design and Analysis of Strategies for the best-of-n Problem

 Springer

Gabriele Valentini
School of Earth and Space Exploration
Arizona State University
Tempe, AZ
USA

ISSN 1860-949X   ISSN 1860-9503   (electronic)
Studies in Computational Intelligence
ISBN 978-3-319-85196-9   ISBN 978-3-319-53609-5   (eBook)
DOI 10.1007/978-3-319-53609-5

Printed on acid-free paper

This Springer imprint is published by Springer Nature
The registered company is Springer International Publishing AG
The registered company address is: Gewerbestrasse 11, 6330 Cham, Switzerland

Foreword

In the book, the author considers an important class of collective decision problems known as "best-of-n problems" where the swarm has to choose among a set of possible options and sets to answer the question "how can we program robots in a swarm so that they can make a collective decision for the best-of-n option?" He proposes a novel modular strategy for the solution of problems in this class whose components are option exploration, opinion dissemination, individual decision-making mechanism, and modulation of positive feedback. He then investigates a few instantiations of the proposed modular strategy to solve specific collective decision problems, both in simulation and with real robots.

This book is an essential reading for any scholar interested in understanding how a large number of simple artificial agents, be there robots or software agents, can collectively make a decision in a self-organized way.

Brussels, Belgium   Marco Dorigo
2016

Preface

The successful deployment of large swarms of robots in the real world is tied to their ability to take decisions autonomously. While decision making is generally conceived as the cognitive ability of individual agents to select a belief based only on their preferences and available information, collective decision making is a decentralized cognitive process, whereby an ensemble of agents gathers, shares, and processes information as a single organism and makes a choice that is not attributable to any of its individuals. Collective decision making can be seen as a means of designing and understanding swarm robotics systems. A principled selection of the rules governing this cognitive process allows the designer to define, shape, and foresee the dynamics of the swarm.

This book has been written to provide a formal understanding of the many ways in which robot swarms can take discrete collective decisions and, based on this understanding, to put forward a principled methodology to design collective decision-making strategies. The content of this book is aimed at scholars with little or no previous knowledge of collective decision making in robot swarms as well as experts in the field of swarm robotics looking for a freestanding reference work on the subject. This book is a result of the author's doctoral studies performed at the Université libre de Bruxelles, and its development is based on the author's Ph.D. dissertation "*The Best-of-n Problem in Robot Swarms.*"

In the first part of this book, we introduce the reader to the topic of self-organized collective decision making and then focus on discrete consensus achievement. We formalize the best-of-n problem, define a taxonomy of its possible variants, and use this framework to survey the swarm robotics literature. Successively, we identify a set of mechanisms that are essential for a swarm to take a successful collective decision in the best-of-n problem: option exploration, opinion dissemination, modulation of positive feedback, and individual decision-making mechanism. By leveraging on this understanding, we put forward a modular and model-driven design methodology that allows the designer to compose collective decision-making strategies and to study their dynamics at different level of abstractions. In the second part of this book, we showcase the proposed design methodology. We compose three different collective decision-making strategies, the Indirect Modulation of Majority-based Decisions, the Direct Modulation of Voter-based Decisions, and the Direct Modulation of Majority-based Decisions, and study their performance using both deterministic and stochastic mathematical models. Our analysis of these strategies focuses on the trade-off between the speed at which a swarm takes a collective decision and its accuracy in selecting the optimal option. In the third part of this book, we show how the designed strategies can be readily applied to different real-world scenarios. We pursue this aim by detailing the results of two series of robot experiments. In the first series, we use a swarm of 100 robots to tackle a site-selection scenario where the objective of the swarm is to select which site has the highest quality. In the second series, we show instead how the same strategies apply to a collective perception scenario whereby the objective of the swarm is to decide which feature of an environment is the most spread. Finally, we conclude with a summary of the scientific contributions of this book and discuss possible future direction of research.

Tempe, AZ, USA   Gabriele Valentini

Contents

Symbols

Symbol	Description	Context
$\mathbb{N}$	Set of natural numbers	Best-of-n
$n \in \mathbb{N}$	Number of options of the best-of-n problem	problem
$\{1, 2, \ldots, n\}$	Set of options indexes for n options	
$\{a, b\}$	Set of options indexes for $n = 2$ options	
ρ_i	Quality of option i	
N	Number of agents in the swarm	Agent controller
D_i	Dissemination state for opinion i	swarm
E_i	Exploration state for opinion i	
g	Average unbiased duration of the dissemination state	
σ^{-1}	Average duration of the exploration state	
r	Interaction range of the agents	
VM	Voter model as individual decision-making mechanism	
MR	Majority rule as individual decision-making mechanism	
G	Size of the group of opinions used in the majority rule	
T_N	Consensus time for a swarm of N agents	Mathematical
E_N	Exit probability for a swarm of N agents	models
d_i	Proportion of agents in the dissemination state D_i	
e_i	Proportion of agents in the exploration state D_i	
p_i	Probability to perceive a neighbor with opinion i	
p_{ij}	Probability to change opinion from i to j as a result of an individual decision	
f^d	Function modeling the direct modulation of positive feedback	
f^i	Function modeling the indirect modulation of positive feedback	
$\gamma^\star$	Fixed point in the form $[d_1^\star, \ldots, d_n^\star, e_1^\star, \ldots, e_n^\star]^T$	
$\mathfrak{D}_i$	Molecule specie for agents in the dissemination state D_i	
$\mathfrak{E}_i$	Molecule specie for agents in the exploration state E_i	
D_i	Number of agents in the dissemination state D_i	

(continued)

Symbol	Description	Context
E_i	Number of agents in the exploration state E_i	Mathematical models
P	Stochastic transition matrix	
Q	Matrix of transition probabilities between transient states	
R	Matrix of transition probabilities from a transient to an absorbing state	
F	Fundamental matrix	
I	Identity matrix	
O	Matrix with all entries set to 0	
ξ	Column vector of all 1s	
τ	Expected number of steps before absorption	
τ_2	Variance of the number of steps before absorption	

Chapter 1
Introduction

We begin this book by motivating the need of a formal understanding of collective decisions in robot swarms. We introduce the objectives of the work described in this book and illustrate the reasoning underlying our research approach. We conclude with an outline of the book's structure where we anticipate the scientific contributions provided in each chapter.

1.1 Motivation and Approach

The ability to undertake a decision-making process is a fundamental property of both natural and artificial systems regardless of their scale, number of components, computational and sensory capabilities. In artificial systems, the paramount importance of decision-making has long been recognized by the robotics and artificial intelligence community as a means necessary to obtain autonomy (Jennings et al. 1998). Systems composed of a single agent are supported by comprehensive theoretical frameworks that include decision trees, Markov decision processes, and reinforcement learning, to name a few. Cooperation and deception strategies between multiple rational decision-makers have been largely explored in game theory, extending the subject of decision-making to a group of two or more agents. Yet, we still lack a comprehensive theory of decision-making in the case of large populations of interacting agents as the ones of robot swarms.

Swarm robotics (Dorigo et al. 2014; Şahin 2005) studies the application of the principles of swarm intelligence (Beni 2005) to design and control large groups of autonomous robots. Swarm robotics aims at designing systems that are scalable in the number of robots and keep performing for increasing swarm sizes; systems that are robust to failure of individual robots thanks to their high redundancy of components and that are flexible to unknown and/or changing environments. In order to meet these objectives, swarm robotics systems are designed following the working

© Springer International Publishing AG 2017
G. Valentini, *Achieving Consensus in Robot Swarms*, Studies in Computational Intelligence 706, DOI 10.1007/978-3-319-53609-5_1

principle of self-organization: "*order through fluctuations*" (Nicolis and Prigogine 1977). Random fluctuations are amplified by positive feedback until the system converges to an ordered state and then dampened by negative feedback to keep the system ordered. In a swarm robotics system, individual agents rely only on partial and local knowledge of their environment and their interactions are governed by simple control rules. Nonetheless, repeated interactions among agents and between agents and the environment allow a swarm to act as an integrated entity capable of performing autonomously.

Swarm robotics has been shown to provide a promising approach for a number of different applications. Robots in a swarm can self-assemble to form complex structures (Klavins 2007; Nagpal 2002; O'Grady et al. 2009; Rubenstein et al. 2014), they can cooperate to perform construction works (Ardiny et al. 2015; Werfel et al. 2014) as well as to collectively transport (Ferrante et al. 2013; Rubenstein et al. 2013; Wilson et al. 2014) or manipulate objects (Ijspeert et al. 2001). The successful accomplishment of these tasks requires the swarm to address high-level cognitive problems. For example, a precondition of a self-assembly application may require the swarm to identify the most suitable shape to form (Christensen et al. 2007); when performing collective construction (Soleymani et al. 2015), the swarm may need to find the most favorable working location (Campo et al. 2010) or the shortest path to efficiently transport materials from a central depot (Montes de Oca et al. 2011). Moreover, complex application scenarios can be tackled by decomposing them into a combination of multiple but simpler tasks (e.g., collective construction can be conveniently decomposed into foraging, transport, and sorting tasks, as suggested by Parker and Zhang 2006). However, in order to undertake this divide and conquer approach, the swarm would need to face and to address several distinct collective decision-making problems.

The development of this book is motivated by the need in swarm robotics for a comprehensive theoretical framework of collective decisions such as the ones described above. We consider collective decision-making problems that demand the swarm to find consensus over which option of a finite set of alternatives offers the most profitable choice for the swarm. We say that this type of cognitive problems requires *discrete consensus achievement*. As we will see in the next chapter, a substantial number of research studies have considered application scenarios requiring discrete consensus achievement. However, the majority of these studies resulted in domain-specific solutions difficult to port to other scenarios. Our objective is to put forward a principled and theoretically grounded methodology to systematically design collective decision-making strategies for discrete consensus achievement problems. Our approach is based on the idea that the solution of many different application scenarios might require the swarm to address the same cognitive problem, i.e., the best-of-n problem. We first study the structure and the variants of the best-of-n problem and then, based on the newly gathered knowledge, we look at the structure of the collective decision-making strategy necessary to a swarm to address a specific problem configuration.

As opposed to the close mimicking of natural systems or to the engineering of domain-specific strategies, we aim to devise a methodology to design collective decision-making strategies that are sufficiently generic to be ported across different applications scenarios. To this end, we decouple the high-level control logic of a strategy from its low-level implementation details which are necessarily tailored for a specific scenario and robotic platform. This approach allows us to identify a minimal set of fundamental processes and their mechanisms that are necessary to implement a decentralized consensus achievement process: *option exploration*, to gather information from the environment; *opinion dissemination*, to spread the gathered information within the swarm; *modulation of positive feedback*, to bias the decision-making process in favor of the best option; and *individual decision-making mechanism*, to allow agents to change their opinions. We build on this understanding to define a modular control structure of a generic collective decision-making strategy and to provide guidelines and constraints for the design of specific modules. This approach allows us to define a generic compartmental model that can be instantiated by the designer to analytically study the macroscopic dynamics of a specific choice of modules. As a result of this endeavor, we support the designer with a model-driven approach that facilitates the selection and comparison of different design choices for the modules of a strategy.

1.2 Outline

The rest of this book is organized in four parts.

In Part I, which consists of Chaps. 2 and 3, we provide the reader with the background necessary to fully appreciate the contributions of this book.

Chapter 2 gives an overview of the literature on collective decision making in robot swarms. First, we organize collective decision-making systems in two categories: consensus achievement and task allocation. Then, we focus on systems designed for consensus achievement problems and further distinguish them between discrete and continuous. Finally, we formalize the framework of the best-of-n problem and show how this framework can be specialized to cover a large number of application scenarios that require discrete consensus achievement by reviewing the swarm robotics literature.

Chapter 3 outlines our modular methodology to design a collective decision-making strategy for the best-of-n problem. We identify the fundamental processes that determine the functioning of a collective decision-making strategy. Successively, we propose a modular structure of a generic strategy that implements these fundamental processes and provide guidelines to design specific modules. Finally, we introduce a general methodology to derive predictive mathematical models and show how this methodology can be instantiated to model a specific combination of modules defining a strategy.

In Part II, which consists of Chaps. 4, 5, and 6, we illustrate the use of the proposed modular design methodology by developing three different collective decision-making strategies, describing them by means of predictive mathematical models, and analyzing their collective dynamics.

Chapter 4 describes the Indirect Modulation of Majority-based Decisions strategy (IMMD). We study the performance of the IMMD strategy numerically by means of macroscopic Monte Carlo simulations and analytically by defining and analyzing an absorbing, time-homogeneous Markov chain that models a system with a finite number of robots.

Chapter 5 introduces the Direct Modulation of Voter-based Decisions strategy (DMVD). We study the performance of the DMVD strategy numerically by means of microscopic multi-agent simulations where robots are represented by mass-less particles moving in a closed environment. Additionally, we perform an analytical study of the proposed strategy by means of two macroscopic mathematical models: a deterministic mean-field model, that consists of a system of ordinary differential equations, and a stochastic model, that consists of a chemical reaction network.

Chapter 6 introduces the Direct Modulation of Majority-based Decisions strategy (DMMD). We study the performance of the DMMD strategy by means of two mathematical models, respectively, a system of ordinary differential equations and a chemical reaction network. The analysis contained in this chapter focuses on the speed versus accuracy trade-off characterizing the DMMD strategy. Additionally, we compare the performance of the DMMD strategy against the one of the DMVD strategy.

In Part III, which consists of Chaps. 7 and 8, we validate the results obtained in the previous part of this book by means of real-robot experiments.

Chapter 7 reports the results of robot experiments performed in a site-selection scenario. We provide an implementation of the DMMD strategy that is tailored for the Kilobot robot and compare the obtained performance with the predictions of the mathematical models defined in Chap. 6.

Chapter 8 reports the results of robot experiments performed in a collective perception scenario. We support the generality of the modular design methodology proposed in this book by providing implementations of both the DMVD strategy and the DMMD strategy that are tailored for the e-puck robot. The analysis presented in this chapter is performed using both physics-based simulations and experiments with real robots.

In Part IV, which consists of Chap. 9 and Appendix A, we conclude this book and we provide the reader with an appendix.

Chapter 9 concludes this book. We summarize the research contributions described in this book and discuss their relevance. Finally, we conclude by suggesting and discussing possible future directions of research.

Appendix A provides a basic background on the formalism of time-homogeneous Markov chains focused on the analysis of absorbing Markov chains.

References

H. Ardiny, S. Witwicki, F. Mondada, Construction automation with autonomous mobile robots: a review, in *2015 3rd RSI International Conference on Robotics and Mechatronics, (ICROM)* (IEEE Press, 2015), pp. 418–424

G. Beni, From swarm intelligence to swarm robotics, ed. By E. Şahin, W.M. Spears. *Swarm Robotics*, vol. 3342. LNCS (Springer, 2005), pp. 1–9

A. Campo, Á. Gutiérrez, S. Nouyan, C. Pinciroli, V. Longchamp, S. Garnier, M. Dorigo, Artificial pheromone for path selection by a foraging swarm of robots. Biol. Cybern. **103**(5), 339–352 (2010)

A.L. Christensen, R. O'grady, M. Dorigo, Morphology control in a multirobot system. IEEE Robot. Autom. Mag. **14**(4), 18–25 (2007)

M. Dorigo, M. Birattari, M. Brambilla, Swarm robotics. Scholarpedia **9**(1), 1463 (2014)

E. Ferrante, M. Brambilla, M. Birattari, M. Dorigo, Socially-mediated negotiation for obstacle avoidance in collective transport, ed. By A. Martinoli, F. Mondada, N. Correll, G. Mermoud, M. Egerstedt, A.M. Hsich, E.L. Parker, K. Støy. *Distributed Autonomous Robotic Systems*, vol. 83. STAR (Springer, 2013), pp. 571–583

A.J. Ijspeert, A. Martinoli, A. Billard, L.M. Gambardella, Collaboration through the exploitation of local interactions in autonomous collective robotics: the stick pulling experiment. Auton. Robots **11**(2), 149–171 (2001)

N.R. Jennings, K. Sycara, M. Wooldridge, A roadmap of agent research and development. Auton. Agents Multi Agent Syst. **1**(1), 7–38 (1998)

E. Klavins, Programmable self-assembly. IEEE Control Syst. **27**(4), 43–56 (2007)

M.A. Montes de Oca, E. Ferrante, A. Scheidler, C. Pinciroli, M. Birattari, M. Dorigo, Majority-rule opinion dynamics with differential latency: a mechanism for self-organized collective decision-making. Swarm Intell. **5**, 305–327 (2011)

R. Nagpal, Programmable self-assembly using biologically-inspired multiagent control, in *Proceedings of the First International Conference on Autonomous Agents and Multiagent Systems, AAMAS'02* (ACM, 2002), pp. 418–425

G. Nicolis, I. Prigogine, *Self-Organization in Nonequilibrium Systems* (Wiley, New York, 1977)

R. O'Grady, A.L. Christensen, M. Dorigo, SWARMORPH: multirobot morphogenesis using directional self-assembly. IEEE Trans. Robot. **25**(3), 738–743 (2009)

C.A.C. Parker, H. Zhang, Collective robotic site preparation. Adapt. Behav. **14**(1), 5–19 (2006)

M. Rubenstein, A. Cabrera, J. Werfel, G. Habibi, J. McLurkin, R. Nagpal, Collective transport of complex objects by simple robots: theory and experiments, ed. By T. Ito, C. Jonker, M. Gini, O. Shehory, *Proceedings of the 12th International Conference on Autonomous Agents and Multiagent Systems, AAMAS '13* (IFAAMAS, 2013), pp. 47–54

M. Rubenstein, A. Cornejo, R. Nagpal, Programmable self-assembly in a thousand-robot swarm. Science **345**(6198), 795–799 (2014)

E. Şahin, Swarm robotics: from sources of inspiration to domains of application, ed. By E. Şahin, W. Spears. *Swarm Robotics*, vol. 3342. LNCS (Springer, 2005), pp. 10–20

T. Soleymani, V. Trianni, M. Bonani, F. Mondada, M. Dorigo, Bio-inspired construction with mobile robots and compliant pockets. Robot. Auton. Syst. **74**, Part B, 340–350 (2015)

J. Werfel, K. Petersen, R. Nagpal, Designing collective behavior in a termite-inspired robot construction team. Science **343**(6172), 754–758 (2014)

S. Wilson, T.P. Pavlic, G.P. Kumar, A. Buffin, S.C. Pratt, S. Berman, Design of ant-inspired stochastic control policies for collective transport by robotic swarms. Swarm Intell. **8**(4), 303–327 (2014)

Part I
Background and Methodology

Chapter 2
Discrete Consensus Achievement in Artificial Systems

Collective decision making refers to the phenomenon whereby a collective of agents makes a choice in a way that, once made, the choice is no longer attributable to any of the individual agents. This phenomenon is widespread across natural and artificial systems and is studied in a number of different disciplines including psychology, biology, and physics, to name a few. In the case of robot swarms, collective decision-making systems are distinguished between systems for *consensus achievement* and systems for *task allocation*. The first category encompasses systems that aim to establish an agreement among agents on a certain matter. The second category deals with systems that aim to allocate agents, i.e., the available workforce, to a set of tasks with the objective to maximize the performance of the collective. In this chapter, we focus on consensus achievement problems. We define the best-of-n problem and a taxonomy of its possible variants. Using this problem-oriented taxonomy, as well as a second taxonomy based on the design methodology, we review the literature of swarm robotics and provide a complete overview of the current state of the art.

2.1 Consensus Achievement

Consensus achievement problems can be further distinguished in two classes depending on the granularity of the choices available to the swarm. When the possible choices of the swarm are finite and countable, we say that the consensus achievement problem is *discrete*. An example of a discrete problem is the selection of the shortest path connecting the entry of a maze with its exit (Szymanski et al. 2006). Alternatively, when the choices of the swarm are infinite and measurable, we say that the consensus achievement problem is *continuous*. An example of a continuous problem is the selection of a common direction of motion by a swarm of agents flocking in a two- or three-dimensional space (Reynolds 1987). Both discrete and continuous consensus achievement problems have already received substantial attention from the scientific community.

© Springer International Publishing AG 2017
G. Valentini, *Achieving Consensus in Robot Swarms*, Studies in Computational Intelligence 706, DOI 10.1007/978-3-319-53609-5_2

Discrete consensus achievement problems have been studied in a number of different contexts. The community of artificial intelligence focused on decision-making approaches for cooperation in teams of agents applying methods from the theory of decentralized partially observable Markov decision processes (Bernstein et al. 2002; Pynadath and Tambe 2002). These approaches, however, rely on sophisticated communication strategies and are suitable only for relatively small teams of agents. Discrete consensus achievement problems have been considered also in the context of the RoboCup soccer competition (Kitano et al. 1997). In this scenario, robots in a team are provided with a predefined set of plays and are required to agree on which play to execute. Different decision-making approaches have been developed to tackle this problem including centralized (Bowling et al. 2004) and decentralized (Kok and Vlassis 2003; Kok et al. 2003) play-selection strategies. Other approaches to consensus achievement over discrete problems have been developed in the context of sensor fusion to perform distributed object classification (Kornienko et al. 2005a, b). Finally, discrete consensus achievement problems are also studied by the community of statistical physics. Example studies include models of collective motion in one-dimensional spaces (Czirók and Vicsek 2000; Czirók et al. 1999; Yates et al. 2009) that describe the marching bands phenomenon of locust swarms (Buhl et al. 2006) as well as models of democratic voting and opinion dynamics (Castellano et al. 2009; Galam 2008).

Continuous consensus achievement problems have been primarily studied in the context of collective motion, that is, flocking (Camazine et al. 2001). Flocking is the phenomenon whereby a collective of agents moves cohesively in a common direction. The selection of a shared direction of motion represents the consensus achievement problem. In swarm robotics, flocking has been studied in the context of both autonomous ground robots (Ferrante et al. 2012; Nembrini et al. 2002; Spears et al. 2004; Turgut et al. 2008) and unmanned aerial vehicles (Hauert et al. 2011; Holland et al. 2005) with a focus on developing algorithms suitable for minimal and unreliable hardware. Apart from flocking, the swarm robotics community focused on spatial aggregation scenarios where robots are required to aggregate in the same region of a continuous space (Garnier et al. 2008; Gauci et al. 2014; Soysal and Şahin 2007; Trianni et al. 2003). The phenomenon of flocking is also studied by the community of statistical physics (Szabó et al. 2006; Vicsek and Zafeiris 2012) with the aim of defining a unifying theory of collective motion that equates several natural systems. A popular example study is provided by the minimalist model of self-driven particles proposed by Vicsek et al. (1995). The community of control theory has intensively studied the problem of consensus achievement (Mesbahi and Egerstedt 2010) with the objective of deriving optimal control strategies and prove their stability. In addition to flocking and tracking (Cao and Ren 2012), the consensus achievement problems studied in control theory include formation control (Ren et al. 2005), agreement on state variables (Hatano and Mesbahi 2005), sensor fusion (Ren and Beard 2008) as well as the selection of motion trajectories (Sartoretti et al. 2014). Finally, continuous consensus achievement problems have been also studied in the context of wireless sensor networks with the aim of developing algorithms for distributed estimation of signals (Schizas et al. 2008a, b).

In the rest of this chapter, we focus on discrete consensus achievement scenarios and we overview a number of research studies that proposed collective decision-making strategies specifically conceived for robot swarms. First, we formally define the best-of-n problem, i.e., a general structure and logic of a decision-making problem that characterizes several application scenarios in swarm robotics. Successively, we review related studies by organizing them in different classes depending on the approach adopted to design the collective decision-making strategy. Finally, we discuss the main differences between the different design approaches.

2.2 The Best-of-n Problem

In the swarm robotics literature, a large number of research studies focused on a relatively few application scenarios whose accomplishment requires the swarm to solve a consensus achievement problem (e.g., the shortest-path problem in foraging scenarios, site-selection in aggregation scenarios). These application scenarios have been primarily tackled separately from each other with an application-oriented perspective that resulted in either the development of domain-specific methodologies or the design of black-box controllers (cf. Sect. 2.3). However, we believe that the consensus achievement problems underlying these application scenarios share a similar logic and structure and that they can be abstracted to a unique framework: the *best-of-n problem*.

From an abstract point of view, the best-of-n problem requires a swarm of agents to make a collective decision over which option, out of n available options, offers the best alternative to satisfy the current needs of the swarm. We use the generic term *options* to abstract domain-specific concepts such as foraging patches, aggregation areas, or traveling paths, to name a few, that are related to particular application scenarios. We refer to the different options of the best-of-n problem using natural numbers, $1, \ldots, n$, and we say that the swarm is required to find the option $i \in \{1, \ldots, n\}$ with highest quality. That is, each option $i \in \{1, \ldots, n\}$ is characterized by an option quality ρ_i. Without loss of generality, we consider the quality of each option i to be normalized in the interval $(0; 1]$ and $\rho_i = 1$ to represent the quality of the best option. Again, we use the term *option quality* as an abstraction to represent domain specific features (e.g., the length of a path, the size of an aggregation spot, the quality of food in a foraging patch).

Given a swarm of N agents, we say that the swarm has found a solution to a particular instance of the best-of-n problem as soon as it makes a *collective decision* for any option $i \in \{1, \ldots, n\}$. A collective decision is represented by the establishment of a *large majority* $M \geq (1 - \delta)N$ of agents that favor the same option i, where δ, $0 \leq \delta \ll 0.5$, represents a tolerance threshold set by the designer. In the boundary case with $\delta = 0$, we say that the swarm has reached a *consensus* decision, i.e., all agents favor the same option i. It is worth noting two aspects of a collective decision. On the one hand, a collective decision must satisfy the property of cohesion (Franks et al. 2013); that is, $\delta \ll 0.5$ implies that the opinions within the swarm are not split over different options of the best-of-n problem. On the other hand, a collective decision inherits the quality of the associated option i and, therefore, it can be optimal, $\rho_i = 1$, or sub-optimal, $\rho_i < 1$.

In general, the quality of an option is a function of the features of the environment, or of characteristics inherent to the swarm (e.g., the number of agents), or a combination of both factors and possibly multiple attributes (Reid et al. 2015). We distinguish between two factor types that determine the quality of a certain option. On the one hand, the option quality can be determined by an *internal preference* of individual agents for specific attributes characterizing an option. For example, when searching for a new nest site, honeybees instinctively favor candidate sites with a certain volume, exposure, and height from the ground (Camazine et al. 1999) regardless of their distance from the current nest location. This type of factors requires individual agents to directly evaluate the attributes of a certain option and to estimate its quality. On the other hand, the option quality can be determined by an existing bias that does not generate internally to individual agents but from certain features of the environment that indirectly influence the behavior of the swarm. We refer to this type of factors as *environmental bias*. For example, when foraging, ants find the shortest traveling path between a pair of locations as a result of pheromone trails being reinforced more often on the shortest path (Goss et al. 1989). Ants do not measure the length of each path individually. However, the length of a path indirectly influences the amount of pheromone laid over the path by the ants. Environmental bias factors can be interpreted as defining the cost of each option; this environmental cost can affect the selection of the best option by the swarm both positively, when higher quality options have lower costs, or negatively, otherwise (cf. Sect. 3.2.2).

In the case in which the option quality is independent of internal agent preferences and of environmental bias, the best-of-n problem reduces to a symmetry-breaking problem (de Vries and Biesmeijer 2002; Hamann et al. 2012). In this case, any of the n available options of the decision-making problem has the same quality and the goal of the swarm is to collectively choose one of the available options. A symmetry-breaking scenario arises also as a special case of the other classes when two or more options have equal and highest quality.

Finally, depending on the considered application scenario, the option quality is either *dynamic* or *static*. That is, the value of ρ_i may be a function of time. This feature is particularly relevant to guide the choices of designers when defining a collective decision-making strategy. When the option quality is static, the decision-making problem is a non-recurring or rarely recurring problem. In this case designers favor strategies that results in consensus decisions (Montes de Oca et al. 2011; Parker and Zhang 2009; Scheidler et al. 2016). Differently, when the option quality is a function of time (Arvin et al. 2014; Parker and Zhang 2010), designers favor strategies that result in a large majority of agents in the swarm favoring the same option without converging to consensus. In this case, the remaining minority of agents that are not aligned with the current swarm decision keep exploring other options and possibly discover new ones. This approach makes the swarm adaptive to changes in the environment (Schmickl et al. 2009b).

2.3 Overview of Current Design Approaches

The efforts of researchers in the last decade produced a vast literature of studies that spans over a number of application scenarios, design approach, and resulting collective decision-making strategies. In this section, we introduce a taxonomy that will be used in the rest of this chapter to review the most important studies performed in the literature of swarm robotics (see Fig. 2.1).

Particular instances of the best-of-n problem have been tackled using both bottom–up and top–down design approaches (Crespi et al. 2008). In the bottom–up approach, the designer develops the robot controller by hand, following a trial and error procedure where the robot controller is iteratively refined until the swarm behavior fulfills the requirements. In the top–down approach, the controller for individual robots is derived directly from a high-level specification of the desired behavior of the swarm by means of automatic techniques, for example, as a result of an optimization process (Bongard 2013; Nolfi and Floreano 2000) or of a compilation process (Werfel et al. 2014).

In the bottom–up approaches (see Sect. 2.4), the robot controller is usually developed by defining different atomic behaviors that are combined together by the designer to obtain a probabilistic-finite state machine. Each behavior of the robot controller is implemented by a set of control rules that determines (i) how a robot works on a certain task and (ii) how it interacts with its neighbor robots. We distinguish collective decision-making strategies designed by means of a bottom–up process in two categories, cf. Fig. 2.1, depending on how the control rules governing the interaction among robots have been defined. In the first category, that we call opinion-based approaches, robots have an explicit internal representation of their favored option, i.e., an opinion, and the role of the designer is to define the control rules that determine how robots share and change their opinions. Opinion-based approaches are used by the designer to tackle directly a consensus achievement problem rather than specific application scenarios. In the second category, that we call ad hoc approaches, we consider research studies where the control rules governing the interaction between robots have been defined by the designer to address

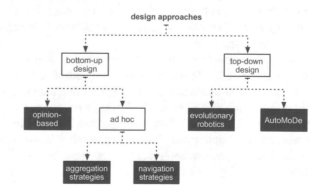

Fig. 2.1 Taxonomy used to review research studies that consider a discrete consensus achievement scenario. Research studies are organized according to their design approach (i.e., bottom–up and top–down) and to how the control rules governing the interaction among robots have been defined

a specific application scenario. As opposed to opinion-based approaches, control strategies belonging to this category are not explicitly designed to solve a consensus achievement problem; nonetheless, their execution by robots of the swarm results in a collective decision. In this category, we consider research studies that focus on the problem of spatial aggregation and on the problem of navigation in unknown environments.

In the top–down approaches (see Sect. 2.5), the robot controller is derived automatically from a high-level description of the desired swarm behavior. We organize research studies adopting a top–down approach in two categories: evolutionary robotics and automatic modular design. Evolutionary robotics (Bongard 2013; Nolfi and Floreano 2000) relies on evolutionary computation methods to obtain a neural network representing the robot controller. As a consequence, this design approach results in black-box controllers. In contrast, automatic modular design (Francesca et al. 2014) relies on optimization processes to combine behaviors chosen from a predefined set and obtain a robot controller that is represented by a probabilistic finite-state machine.

2.4 Bottom–Up Design Approaches

In this section, we consider research studies that developed collective decision-making strategies using a bottom–up design approach.

2.4.1 Opinion-Based Approaches

A large amount of research work has focused on the design of collective decision-making strategies characterized by robots having an explicit representation of their opinions. We refer to these collective decision-making strategies as opinion-based approaches. Using this design approach, robots are required to perform *explicit information transfer* (Ferrante 2013), i.e., to purposely transmit information represented by their opinion to their neighbors. As a consequence, a collective decision-making strategy developed using an opinion-based approach requires robots to have communication capabilities.

One of the first research studies developed with an opinion-based approach is that of Wessnitzer and Melhuish (2003). The authors considered a scenario in which a swarm of robots needs to capture two preys that are moving in the environment (i.e., the option of a best-of-2 problem). To do so, robots are required to collectively choose which prey to hunt first. The authors proposed a collective decision-making strategy based on the majority rule. Initially, each robot favors a prey chosen at random. At each time step, robots apply the majority rule over their neighborhood in order to reconsider and possibly change their opinions. In this case study, the two options of the decision-making problem are characterized by the same quality and thus the decision-making problem requires the swarm to break this symmetry.

Parker and Zhang (2009) developed a collective decision-making strategy by taking inspiration from the house-hunting behavior of social insects (Franks et al. 2002). The authors considered a site-selection scenario where a swarm of robots is required to discriminate between two illuminated sites (i.e., the options of a best-of-2 problem) based on their level of brightness (i.e., option quality defined by an internal preference factor). The proposed control strategy is characterized by three phases. Initially, robots are in the search phase either exploring the environment or waiting in a idle state. Upon discovery of a site and estimating its quality, a robot transits to the deliberation phase. During the deliberation phase, a robot recruits other robots in the search phase by repeatedly sending recruitment messages. The frequency of these messages is proportional to the option quality. Meanwhile, robots estimate the popularity of their favored option and use this information to test if a quorum has been reached. Upon detection of a quorum, robots enter the commitment phase and eventually relocate to the chosen site. The strategy proposed by Parker and Zhang builds on a direct recruitment and a quorum-sensing mechanism inspired by the house-hunting behavior of ants of the *Temnothorax* species. Later, Parker and Zhang (2011) considered a simplified version of this strategy and proposed a rate equation model to study its convergence properties.

Parker and Zhang (2010) proposed a collective decision-making strategy for unary decisions and applied it to the task sequencing problem. In the task sequencing problem, a swarm of robots needs to work sequentially on different tasks. The robots are required to collectively agree on the completion of a certain task prior to begin working on the next task. The task sequencing problem is a best-of-2 problem whose options (i.e., "task complete" or "task incomplete") are characterized by dynamic qualities (i.e., the level of completeness of a task changes over time). The authors proposed a quorum-sensing strategy to address this problem. Robots working on the current task monitor its level of completion (and, therefore, the option quality is due to an internal preference factor); when a robot recognizes the completion of the task, it enters the deliberation phase during which it asks its neighbors if they recognized too the completion of the task. Once a deliberating robot perceives a certain number of neighbors in the deliberation phase (i.e., the quorum), it moves to the committed phase during which it sends commit messages to inform neighbor robots about the completion of the current task. Robots in the deliberation phase that receive a commit message enter the committed phases and responds with an acknowledgement message. Committed robots measure the time passed since the last received acknowledgement and, after a certain time, they begin working on the next task.

Rather than mimicking biological systems, Montes de Oca et al. (2011) took advantage of the theoretical framework developed in the field of opinion dynamics (Krapivsky and Redner 2003). The authors extended the concept of latent voters introduced by Lambiotte et al. (2009) (i.e., voters stop participating to the decision-making process for a stochastic amount of time after changing opinion) and proposed a collective decision-making strategy referred to as *majority rule with differential*

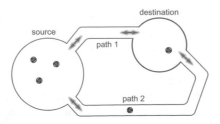

Fig. 2.2 Schematic illustration of the double-bridge problem (Goss et al. 1989). A source and a destination areas are connected by 2 paths: path 2 (*blue arrow*) whose length is approximately twice longer than path 1 (*red arrow*). The figure also shows a swarm of 5 robots (*green circles*), 3 robots with opinion 1 (*red triangle*) and 2 robots with opinion 2 (*blue triangle*)

latency. The proposed strategy is applied to a scenario inspired by the popular *double-bridge* problem (Goss et al. 1989). Robots in a swarm need to transport objects between two locations (i.e., source and destination in Fig. 2.2) connected by two paths of different length (i.e., option quality). Objects are too heavy to be transported by single robots and require a team of 3 robots. During the collective decision-making process, robots repeatedly form teams at the source location. Within a team, robots share with each other their opinion for their favored path and then apply the major-ity rule (Galam 2008) to determine which path the team should traverse. Then, the team travels back-and-forth along the chosen path before dismantling once back in the source location. Using this strategy, robots do not measure the length of each path; in contrast, the length of a path indirectly biases the frequency of participa-tion of robots to the decision-making process taking place in the source location (i.e., the option quality is defined by an environmental bias). The majority rule with differential latency has been the subject of an extensive theoretical analysis that includes deterministic macroscopic models (Montes de Oca et al. 2011), master equations (Scheidler 2011), statistical model checking (Massink et al. 2013), and Markov chains (see Chap. 4).

The same foraging scenario investigated by Montes de Oca et al. (see Fig. 2.2) has been the subject of other research studies. In Brutschy et al. (2012); Scheidler et al. (2016), the authors extended the control structure underlying the majority rule with differential latency introducing the k-unanimity rule—a novel decision-making mechanism for individual robots. Instead of forming teams and applying the majority rule within each team, robots have a memory of size k where they store the opinions of other robots as they encounter them. A robot using the k-unanimity rule changes its current opinion in favor of a different option only after consecutively encountering k other robots all favoring that other option. The primary benefit of the k-unanimity rule is that it allows the designer to adjust the speed and the accuracy of the collective decision-making strategy by means of the parameter k (Scheidler et al. 2016). The authors studied the dynamics of the k-unanimity rule analytically when applied to decision-making problems with up to $n = 3$ options using a deterministic macroscopic model and a master equation.

Reina et al. (2014, 2015a,b) proposed a collective decision-making strategy inspired by theoretical studies that unify the decision-making behavior of social insects with that of neurons in vertebrate brains (Marshall et al. 2009; Pais et al. 2013; Seeley et al. 2012). The authors considered the problem of finding the shortest path connecting a pair of locations in the environment. In their strategy, robots can be either uncommitted, i.e., without any opinion favoring a particular option, or committed to a certain option, i.e., with an opinion. Uncommitted robots might discover new options in which case they become committed to the discovered option. Committed robots can recruit other robots that have not yet an opinion (i.e., direct recruitment); inhibit the opinion of robots committed to a different option making them become uncommitted (i.e., cross-inhibition); or abandon their current opinion and become uncommitted (i.e., abandonment). In Reina et al. (2014, 2015a), the authors studied the proposed strategy in a foraging scenario with two alternative foraging patches (i.e., the option of a best-of-2 problem); the quality of each option is determined by their distance from a central retrieval area which indirectly influence the behavior of the swarm (i.e., environmental bias). In Reina et al. (2015b), the authors studied the proposed strategy in a different setup: foraging patches are characterized by a quality that the robot can measure (i.e., the internal preference factor) and are positioned at different distances in a way that the best foraging patch is the farthest (i.e., an environmental bias factor that influences negatively the decision-making process). The proposed strategy is supported by both deterministic and stochastic mathematical models (i.e., ordinary differential equations and chemical reaction networks) that link the microscopic parameters of the system to the macroscopic dynamics of the swarm.

2.4.2 Ad Hoc Approaches

In this section, we consider a number of research studies that resulted in the development of control strategies for specific application scenarios, that is, spatial aggregation and navigation in unknown environments. As opposed to opinion-based approaches, the objective of the designers of these control strategies is not to tackle a consensus achievement problem but to address a specific need of the swarm (i.e., aggregation or navigation). Nonetheless, the control strategies reviewed in this section provide a swarm of robots with collective decision-making capabilities.

2.4.2.1 Aggregation-Based Approaches

Aggregation-based approaches are control strategies that make the robots of the swarm aggregate in a common region of the environment forming a cohesive cluster. The opinion of a robot is represented implicitly by its position in space. The primary advantage of an aggregation strategy is represented by the fact that the information regarding a robot opinion can be implicitly transferred to nearby robots without the

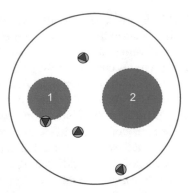

Fig. 2.3 Schematic illustration of the aggregation scenario considered in Campo et al. (2010a); Garnier et al. (2009). The environment is represented by a circular arena enclosed by a wall (*blue line*). The arena is provided with 2 shelters of different size: shelter 1 (*red circle*) has an area approximately twice as smaller than that of shelter 2 (*blue circle*). The figure also shows a swarm of 4 robots (*green circles with purple triangles*) wondering around the environment

need of communication (Ferrante 2013; Sumpter 2010, Chap. 3). Implicit information transfer can be implemented, for example, by means of neighbors observation.

Garnier et al. (2009) considered a behavioral model of self-organized aggregation and studied the emergence of collective decisions. The authors consider an aggregation scenario, cf. Fig. 2.3, where robots are presented with two shelters (i.e., the options of a best-of-2 problem) of different area (i.e., option quality) and are required to select one shelter under which the swarm should aggregate. The proposed control strategy is inspired by the behavior of young larvae of the German cockroach, *Blattella germanica*, (Jeanson et al. 2003, 2005). Robots explore their environment by executing a correlated random walk. When a robot detects the boundary of the arena, it pauses the execution of the random walk and begins the execution of a wall-following behavior. The wall-following behavior is performed for an exponentially distributed period of time after which the robot turns randomly towards the center of the arena. When encountering a shelter, the robot decides whether to stop or not as well as whether to stop for a short or a long period of time as a function of the number of nearby neighbors. Given the number of perceived neighbors, this function returns the probability for a robot to stop and its value has been tuned by the designer to favor the selection of the shelter of bigger area (i.e., shelter 2 in Fig. 2.3). Garnier et al. (2009) studied the proposed strategies in two different setups. In the first setup, the aggregation problem requires to break the symmetry between two shelters of equal size. In the second setup, one shelter is larger than the other. The option quality is determined by an internal preference factor (i.e., the number of perceived neighbors which carries information on the shelter size) and an environmental bias factor (i.e., the shelter size, on the grounds that larger shelters are easier to discover by robots and are of lesser cost). Correll and Martinoli (2011) studied this collective behavior with both Markov chains and difference equations and showed that a collective decision arises only when robots move faster than a minimum speed and are characterized by a sufficiently large communication range.

Campo et al. (2010a) considered the same aggregation scenario of (Garnier et al. 2009) and developed a control strategy taking inspiration from theoretical studies of the aggregation behaviors of cockroaches (Amé et al. 2006). In their strategy the robot controller is composed of 3 phases: exploration, stay under a shelter, and move back to the shelter. Initially, the robots explore the environment by performing a random walk. Once a robot discovers a shelter, it moves randomly within the shelter's area and estimates the density of other robots therein. If during this phase, a robot accidentally exits the shelter, it performs a U-turn aimed at reentering the original shelter. Differently from Garnier et al. (2009), the robots directly decide whether to stay under a shelter or to leave and return to the exploration phase. This decision is stochastic and the probability to leave the shelter is given by a sigmoid function of the estimated density of robots under the shelter. Contrarily to Garnier et al. (2009), the authors tuned the parameters of the sigmoid function with the aim to favor the selection of the smallest shelter that can host the entire swarm (e.g., shelter 1 in Fig. 2.3). As a consequence, the option quality is still determined by a combination of internal preference and environmental bias factors as in Garnier et al. (2009) but, this time, the environmental bias factor (i.e., the size of the shelter) hampers the discovery of the best option (shelter 1 in Fig. 2.3 by the robots of the swarm). A similar aggregation strategy was proposed later by Brambilla et al. (2014) and studied in a binary symmetry-breaking setup. Differently from the sigmoid function used in Campo et al. (2010a), the authors considered a linear function of the number of neighbors to determine the probability with which a robot decides whether to leave a shelter or not.

Kernbach et al. took inspiration from the thermotactic aggregation behavior of young honeybees, *Apis mellifera* L., (Grodzicki and Caputa 2005), and proposed the BEECLUST algorithm (Kernbach et al. 2009; Schmickl et al. 2009b). The goal of a swarm executing the BEECLUST algorithm is to aggregate around the brightest spot in the environment. For this purpose, a robot moves randomly in the environment; upon encountering another robot, the robot stops moving and measures the local intensity of the ambient light. After waiting for a period of time proportional to the measured light, the robot resumes a random walk. In Schmickl et al. (2009b), the authors studied the BEECLUST algorithm in a setup characterized by two spots (i.e., the options of a best-of-2 problem) of different brightness. The option quality is defined by an internal preference factor (i.e., the brightness measured by each robot) and is also positively influenced by an environmental bias factor (i.e., brighter spots are also characterized by a bigger area which make them easier to discover by the robots of the swarm). Later, Hamann et al. (2012) studied the BEECLUST algorithm in a binary symmetry-breaking setup where both spots are characterized by the same level of brightness. The BEECLUST algorithm has been subject of an extensive theoretical analysis that includes both spatial and non-spatial macroscopic models (Hamann 2013; Hamann et al. 2012; Hereford 2010; Schmickl et al. 2009a). While the resulting decision-making process is robust, it is particularly difficult to model due to the complex dynamics of cluster formation and cluster breakup (Hamann et al. 2012).

More recently, Arvin et al. (2012, 2014) extended the original BEECLUST algorithm by means of a fuzzy controller. In the original BEECLUST algorithm, after the expiration of the waiting period, a robot chooses randomly a new direction of motion. Contrarily, using the extension proposed by Arvin et al., the new direction of motion is determined using a fuzzy controller that maps the magnitude and the bearing of the input signal (in their case, a sound signal) to one out of five predetermined directions of motion (i.e., left, slightly-left, straight, slightly-right, right). The authors studied the extended version of the BEECLUST algorithm in a dynamic, binary decision-making problem defined by two aggregation areas; each area is identified by a sound emitter and the sound magnitudes of the two areas are different and vary over time. As in Schmickl et al. (2009b) where the option quality was determined by the level of brightness, the size of each aggregation area is proportional to the magnitude of the emitted sound. Consequently, the option quality is determined by an internal preference factor and is facilitated by an environmental bias factor. This extension has been shown to improve the aggregation performance of the BEECLUST algorithm (i.e., clusters last for a longer period of time) as well as its robustness to noisy perceptions of the environment.

Mermoud et al. (2010) considered an application scenario where robots of the swarm are required to monitor a certain environment, searching and destroying undesirable artifacts (e.g., pathogens, pollution). Specifically, artifacts correspond to colored spots that are projected on the surface of the arena and can be of two types: "good" or "bad". The author proposed an aggregation-based strategy that allows robots to collectively perceive the type of a spot and to destroy those spots that have been perceived as bad while safeguarding good spots. Each robot explores the environment by performing a random walk and avoiding obstacles. Once a robot enters a spot, it measure the light intensity to determine the type of the spot. Successively, the robot moves inside the spot area until it detects a border; at this point, the robot decides with a probability that depends on the estimated spot type whether to leave the spot or to remain inside it by performing a U-turn. Within the spot, a robot stops moving and starts to form an aggregate as soon as it perceives one or more other robots evaluating the same spot. When the aggregate reaches a certain size (which is predefined by the experimenter), the spot is collaboratively destroyed and robots resume the exploration of the environment. The achievement of consensus is detected using an external tracking infrastructure which also emulates the destruction of the spot. The scenario proposed by Mermoud et al. corresponds to an infinite series of best-of-2 decision-making problem (i.e., one for each spot) that are tackled in parallel by different subsets of agents of the swarm (i.e., different robot aggregates). The quality of each spot is determined by an internal preference factor that is represented by the measured light. The proposed strategy has been derived following a bottom–up, multi-level modeling methodology that encompasses physics-based simulations, chemical reaction networks, and continuous ODE approximation (Mermoud et al. 2010, 2014).

2.4.2.2 Navigation-Based Approaches

Navigation-based approaches are control strategies that allow a swarm of robots to navigate an environment towards one or more regions of interest. Navigation algorithms have been extensively studied in the swarm robotics literature. However, not all of them provide a swarm with collective decision-making capabilities. For examples, navigation algorithms based on hop-count strategies have been proposed to find the shortest-path connecting a pair of locations (Payton et al. 2001; Szymanski et al. 2006). However, these strategies are incapable of selecting a unique path when there are two or more paths with equal length and thus fail to make a collective decision (Campo et al. 2010b). From a broader perspective, navigation-based approaches include also flocking whereby robots have to agree on a common direction of motion (Ferrante et al. 2012; Nembrini et al. 2002; Spears et al. 2004; Turgut et al. 2008). However, as discussed in the introduction of this chapter, these control strategies are generally studied in experimental setups corresponding to continuous consensus achievement problems (i.e., best-of-∞). In the following, we consider navigation algorithms applied to discrete consensus achievement problems.

Schmickl and Crailsheim took inspiration from the trophallactic behavior of honeybee swarms, *Apis mellifera* L. (Camazine et al. 1998; Korst and Velthuis 1982), and proposed a virtual gradient and navigation strategy that provides a swarm of robots with collective decision-making capabilities. *Trophallaxis* refers to the direct, mouth-to-mouth exchange of food between two honeybees (or other social insects). The authors studied the proposed strategy in a binary aggregation scenario with two spots of different size (Schmickl et al. 2007) and in a foraging scenario reminiscent of the double-bridge problem (Schmickl and Crailsheim 2006, 2008). Robots explore their environment searching for resources (i.e., aggregation spots, foraging patches). Once a robot finds a resource, it loads a certain amount of virtual nectar. As the robot moves in the environment, it spreads and receives virtual nectar to and from other neighboring robots. This behavior allows robots to create a virtual gradient of nectar that can be used by robots to navigate back and forth a pair of locations following the shortest of two paths or to orient towards the largest of two aggregation areas. In both cases, the quality of each option is determined solely by an environmental bias factor (i.e., the length of a path and the size of the aggregation area) which influences the rate of diffusion of virtual nectar. This trophallaxis-inspired strategy has been studied later using models of Brownian motion (Hamann 2010; Hamann and Wörn 2008). The authors defined both a Langevin equation (i.e., a microscopic model) to describe the motion of an individual agent and a Fokker–Planck equation (i.e., a macroscopic model) to model the motion of the entire swarm finding a good qualitative agreement with simulated dynamics of the trophallaxis-inspired strategy.

Garnier et al. (2007) studied a robot control strategy that closely mimics the pheromone-laying behavior characterizing foraging in many ant species (Deneubourg and Goss 1989; Goss et al. 1989). The authors considered a foraging scenario similar to the double-bridge problem where two areas are connected by a pair of paths of equal length (i.e., the options of a best-of-2 symmetry-breaking problem). During robot experiments, pheromone is emulated by means of an external tracking infrastructure interfaced with a light projector that manages both the laying of pheromone and its evaporation. The robots can perceive pheromone trails by means of a pair of light sensors and can recognize the two target areas by means of IR beacons. In the absence of a trail, a robot moves randomly in the environment avoiding obstacles. When perceiving a trail, the robot starts following the trail and depositing pheromone which evaporates with an exponential decay. In their study, the authors show that using this strategy the robots of a swarm are capable to make a consensus decision for one of the two paths. However, the implementation of pheromone-inspired mechanisms on a robotic platform (Fujisawa et al. 2014) still represents a challenge with current technologies which prevents its employment in real-world robotic applications.

Campo et al. (2010b) proposed a solution to the above limitations of pheromone-inspired mechanisms for the case of chain-based navigation systems. In their research work, the robots of the swarm form a pair of chains leading to 2 different locations. Each chain identifies a path and each path has different length (i.e., 2 options with quality defined by an environmental bias factor). The authors proposed a collective decision-making strategy to select the shortest of the 2 paths that is based on virtual pheromones. Robots in a chain can communicate with their 2 immediate neighbors forming a communication network. Virtual ants navigate through the network and lay virtual pheromone eventually leading to the identification and selection of the shortest path.

Gutiérrez et al. (2009) proposed a navigation strategy called *social odometry* that allows a robot of a swarm to keep an estimate of its current location with respect to a certain area of interest. A robot has an estimate of its current location and a measure of confidence about its belief that decreases with the traveled distance. Upon encountering a neighboring robot, they both exchange their location estimates and confidence measures. Successively, each of the two robots updates its current location estimate by averaging its current location with that of its neighbor weighted by the respective measures of confidence. Using social odometry, Gutiérrez et al. (2010) studied a foraging scenario characterized by two foraging patches (i.e., the options of a best-of-2 problem) positioned at different distance (i.e., option quality) from a central retrieval area. The authors find that the weighted mean underlying social odometry favors the selection by the swarm of the closest foraging patch due to the fact that robots traveling to that patch have higher confidence in their location estimates. In this strategy, the option quality is determined by a combination of an internal preference factor with an environmental bias factor. The internal preference is represented by the level of confidence because it is derived by each robot from its own measured distance. The environmental bias is represented by the distance of a patch from the retrieval area because patches that are closer to the retrieval area are easier to discover by robots and are therefore of lesser cost. Due to the presence of noise, social odometry allows a swarm of robots to find consensus on a common foraging patch also in a symmetric setup where the two patches are positioned at the same distance from the retrieval area.

2.5 Top–Down Design Approaches

In this section, we consider research studies that developed collective decision-making strategies using a top–down design approach. All research studies reviewed below make use of automatic optimization approaches to design robot controllers for specific application scenarios.

2.5.1 Evolutionary Robotics

As for most collective behaviors studied in swarm robotics (Brambilla et al. 2013), collective decision-making systems have been also developed by means of automatic design approaches. The typical automatic design approach is represented by evolutionary robotics (Bongard 2013; Nolfi and Floreano 2000) where optimization methods from evolutionary computation (Back et al. 1997) are used to evolve a population of agent controllers following the Darwinian principles of recombination, mutation, and natural selection. Generally, the individual robot controller is represented by an artificial neural network that maps the sensory perceptions of a robot (i.e., input of the neural network) to appropriate actions of its actuators (i.e., output of the neural network). The parameters of the neural network are evolved to tackle a specific application scenario by opportunely defining a fitness function on a per-case base; the fitness function is then used to evaluate the quality of each controller and to drive the evolutionary optimization process.

Evolutionary robotics has been successfully applied to address a number of collective decision-making scenarios. Trianni and Dorigo (2005) evolved a collective behavior that allows a swarm of physically-connected robots to discriminate the type of holes present on the arena surface based on their perceived width and to decide weather to cross the hole (i.e., the hole is sufficiently narrow to be safely crossed) or to avoid it by changing the motion direction (i.e., the hole is too risky to cross). Similarly, Trianni et al. (2007) considered a collective decision-making scenario where a swarm of robots need to collectively evaluate the surrounding environment and determine weather there are physical obstacles that requires cooperation in the form of a self-assembly or, alternatively, if robots can escape obstacles independently of each other.

Francesca et al. (2012, 2014) applied methods from evolutionary robotics to a binary aggregation scenario similar to that studied in Campo et al. (2010a); Garnier et al. (2008, 2009) but with shelters of equal size (i.e., a symmetry-breaking problem). The authors compared the performance of the evolved controller with theoretical predictions of existing mathematical models (Amé et al. 2006); however, their results show a poor agreement between the two models due to the fact that artificial evolution was capable to exploit specific features (e.g., geometric symmetries) present in the simulated environment.

Evolutionary robotics can be successfully applied to the design of collective decision-making systems. However, its use as a design approach suffers of several drawbacks. For example, artificial evolution is a computationally intensive process and the designer is required to perform it for each specific scenario. Artificial evolution may suffer from over-fitting whereby a successfully evolved controller performs well in simulation but poorly on real robots. This phenomenon is also known as the reality gap (Jakobi et al. 1995; Koos et al. 2013). Moreover, artificial evolution does not provide guarantees on the optimality of the resulting robot controller (Bongard 2013). The robot controller, being ultimately a black-box model, is difficult to model and analyze mathematically (Francesca et al. 2012). As a consequence, in general the designer cannot maintain and improve the designed solutions (Matarić and Cliff 1996; Trianni and Nolfi 2011).

2.5.2 Automatic Modular Design

More recently, Francesca et al. (2014) proposed an automatic design method, called AutoMoDe, that provides a white-box alternative to evolutionary robotics. The robot controllers designed using AutoMoDe are behavior-based and have the form of a probabilistic finite-state machine. Robot controllers are obtained by combining a set of predefined modules (e.g., random walk, phototaxis) using an optimization process that, similarly to evolutionary robotics, is driven by a fitness function defined by the designer for each specific scenario.

Using AutoMoDe, Francesca et al. (2014) designed an aggregation strategy for the same scenario as in Campo et al. (2010a); Garnier et al. (2008, 2009). In their experimental setup, the collective decision-making problem corresponds to a binary symmetry-breaking scenario where the swarm needs to select one of two equally good aggregation spots. The resulting robot controller proceeds as follow. A robot starts in the attraction state in which its goal is to get close to other robots. When perceiving an aggregation spot, the robot stops moving. Once stopped, the robot has a fixed probability for unit of time to return to the attraction state and start moving again. Additionally, the robot may transit to the attraction state in the case in which it has been pushed out of the aggregation spot by other robots.

2.6 Discussion

In this chapter, we introduced the reader to several aspects of collective decision making. We followed Brambilla et al. (2013) and organized collective decisions in consensus achievement and task allocation (Gerkey and Matarić 2004) depending on the purpose of the swarm. We showed how decision-making problems requiring the achievement of consensus can be further distinguished in two classes, discrete and continuous, depending on the granularity of the available options. For the case of discrete consensus achievement, we formally defined the structure of the best-of-n problem and showed how this general framework covers a large number of specific application scenarios. Finally, we reviewed the principal research contributions in swarm robotics that focus on discrete consensus achievement problems. We divided our literature review in two parts, bottom–up (see Sect. 2.4) and top–down (see Sect. 2.5) design approaches. For each part, we further distinguished collective decision-making systems and obtained five different categories: opinion-based, aggregation-based, navigation-based, evolutionary robotics, and automatic modular design.

Aggregation-based approaches to collective decision making have the advantage of functioning without the need of communication by exploiting implicit information transfer. However, aggregation as a means of communicating one own opinion provides a viable solution only when the options of the best-of-n problem are clearly separated in space from each other. Similarly, navigation-based approaches can be applied only to scenarios in which the discrete consensus achievement problem requires the swarm to find the shortest-path connecting different locations. In contrast, automatic design approaches as evolutionary robotics and automatic modular design have the potential to be applied to a larger set of consensus achievement scenarios. Evolutionary robotics, however, might suffer from the reality-gap between simulated and real robots. Moreover, it is difficult to derive predictive mathematical models for systems designed using artificial evolution. This latter limitation might also affect automatic modular design depending on the complexity of the resulting probabilistic finite-state machines. Opinion-based approaches offer a more general design methodology that can be applied to different application scenarios. This higher level of generality, however, requires explicit information transfer and can be obtained only at the cost of robot-to-robot communication.

As introduced in Sect. 2.2, the definition of quality of an option in the best-of-n problem depends on the specific application scenario. Nonetheless, we showed that the option quality can be determined by a combination of factors of two types, internal preference of individual agents and environmental bias indirectly affecting the behavior of the swarm. Figure 2.4 illustrates how different combinations of internal

Fig. 2.4 Classification of consensus achievement scenarios corresponding to the best-of-n problem. The schema illustrates how different combinations of internal preference factors and environmental bias factors influences the best option of the decision-making problem

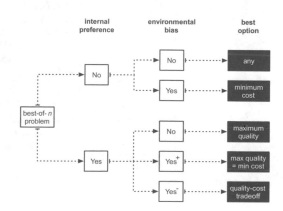

preference and environmental bias factors determine the best option of the best-of-n problem. When the option quality is independent of internal agent preferences and of environmental bias, the best-of-n problem reduces to a symmetry-breaking problem. When the option quality is independent of internal agent preferences and is solely subject to environmental bias, the decision-making problem reduces to finding the option of minimum cost and can be tackled using collective decision-making strategies that do not require agents to directly measure the quality of each option. In the opposite case, i.e., when the option quality depends only on internal preference factors, the best option corresponds to that with highest quality as directly measured by individual agents. Finally, when both factor types coexist, we distinguish between scenarios where the relation between environmental bias and internal preference factors is positive and scenarios in which it is negative. In the first case the best option corresponds to that with highest quality as measured by individual agents which has also minimum cost. In the second case, environmental bias factors influence negatively the option quality and the decision-making problem requires a compromise between option quality and cost.

The taxonomy illustrated in Fig. 2.4 provides us with different means to interpret the swarm robotics literature. We conclude this chapter by reorganizing the research studies reviewed in Sects. 2.4 and 2.5 according to this new taxonomy as shown in Table 2.1. We distinguish research studies in five different categories determined by the specific coupling between internal preference factors and environmental bias

Table 2.1 Classification of the swarm robotics literature according to the combination of factors that determines the quality of the options of the best-of-n problem

Internal preference	Environmental bias	Research lines/studies
No	No	**i.** Wessnitzer and Melhuish (2003) **ii.** Garnier et al. (2007) **iii.** Garnier et al. (2009), Brambilla et al. (2014) **iv.** Hamann et al. (2012), Hamann (2013) **v.** Francesca et al. (2012, 2014)
	Yes	**i.** Hamann (2010); Hamann and Wörn (2008); Schmickl and Crailsheim (2006, 2008); Schmickl et al. (2007) **ii.** Campo et al. (2010b) **iii.** Massink et al. (2013); Montes de Oca et al. (2011); Scheidler (2011) **iv.** Brutschy et al. (2012); Scheidler et al. (2016) **v.** Reina et al. (2014, 2015a)
Yes	No	**i.** Parker and Zhang (2009, 2010, 2011) **ii.** Mermoud et al. (2010, 2014)
	Yes, positive bias	**i.** Garnier et al. (2009) **ii.** Schmickl et al. (2009a, b), Arvin et al. (2012, 2014) **iii.** Gutiérrez et al. (2010)
	Yes, negative bias	**i.** Campo et al. (2010a) **ii.** Reina et al. (2015b)

factors that shapes the best-of-n problem. For each category, we further group the literature in separate lines of research, where each line of research is centered around a particular collective decision-making strategy. In our endeavor, we could not assign some research studies based on evolutionary robotics to a distinct category owing to the difficulty of understanding the precise functioning of the underlying neural networks. The first two categories in Table 2.1, namely, symmetry-breaking problems and problems where the option quality is defined only by environmental bias factors, are covered by a large number of research studies. Moreover, these studies are distributed in five separate lines of research for each category. Differently, the swarm robotics literature has considered a significantly smaller number of studies that focused on the best-of-n problem in the case in which the option quality is determined by an internal preference factor. The majority of these studies considered application scenarios where the option quality is either independent of environmental bias factors or it is positively influenced by them. Note that collective decision-making strategies developed for the former of these two categories directly apply to the latter due to the positive influence of environmental bias factors. The last category, i.e., research studies considering application scenarios where the internal agent preferences are negatively influenced by environmental biases, is the less developed area of study in the literature of discrete consensus achievement with only two research contributions. From the perspective of the designer, this category represents application scenarios with the highest level of complexity that require design solutions able to compensate the negative influence of environmental bias.

References

J.-M. Amé, J. Halloy, C. Rivault, C. Detrain, J.-L. Deneubourg, Collegial decision making based on social amplification leads to optimal group formation. Proc. Natl. Acad. Sci. **103**(15), 5835–5840 (2006)

F. Arvin, A.E. Turgut, S. Yue, Fuzzy-based aggregation with a mobile robot swarm, ed. By M. Dorigo, M. Birattari, C. Blum, A.L. Christensen, A.P. Engelbrecht, R. Groß, T. Stützle. *Swarm Intelligence*, vol. 7461. LNCS (Springer, 2012), pp. 346–347

F. Arvin, A.E. Turgut, F. Bazyari, K.B. Arikan, N. Bellotto, S. Yue, Cue-based aggregation with a mobile robot swarm: a novel fuzzy-based method. Adapt. Behav. **22**(3), 189–206 (2014)

T. Back, D.B. Fogel, Z. Michalewicz (eds.), *Handbook of Evolutionary Computation* (IOP Publishing Ltd., Bristol, 1997)

D.S. Bernstein, R. Givan, N. Immerman, S. Zilberstein, The complexity of decentralized control of markov decision processes. Math. Oper. Res. **27**(4), 819–840 (2002)

J.C. Bongard, Evolutionary robotics. Commun. ACM **56**(8), 74–83 (2013)

M.H. Bowling, B. Browning, M.M. Veloso, Plays as effective multiagent plans enabling opponent-adaptive play selection, in *Proceedings of the 14th International Conference on Automated Planning & Scheduling* (AAAI Press, 2004), pp. 376–383

M. Brambilla, E. Ferrante, M. Birattari, M. Dorigo, Swarm robotics: a review from the swarm engineering perspective. Swarm Intell. **7**(1), 1–41 (2013)

M. Brambilla, A. Brutschy, M. Dorigo, M. Birattari, Property-driven design for robot swarms: a design method based on prescriptive modeling and model checking. ACM Trans. Auton. Adapt. Syst. **9**(4), 17:1–17:28 (2014)

A. Brutschy, A. Scheidler, E. Ferrante, M. Dorigo, M. Birattari, "Can ants inspire robots?" Self-organized decision making in robotic swarms, in *2012 IEEE/RSJ International Conference on Intelligent Robots and Systems (IROS)* (IEEE Press, 2012), pp. 4272–4273

J. Buhl, D.J.T. Sumpter, I.D. Couzin, J.J. Hale, E. Despland, E.R. Miller, S.J. Simpson, From disorder to order in marching locusts. Science **312**(5778), 1402–1406 (2006)

S. Camazine, K. Crailsheim, N. Hrassnigg, G.E. Robinson, B. Leonhard, H. Kropiunigg, Protein trophallaxis and the regulation of pollen foraging by honey bees (Apis mellifera l.). Apidologie **29**(1–2), 113–126 (1998)

S. Camazine, P.K. Visscher, J. Finley, R.S. Vetter, House-hunting by honey bee swarms: collective decisions and individual behaviors. Insectes Sociaux **46**(4), 348–360 (1999)

S. Camazine, J.-L. Deneubourg, N.R. Franks, J. Sneyd, G. Theraulaz, E. Bonabeau, *Self-Organization in Biological Systems* (Princeton University Press, Princeton, 2001)

A. Campo, S. Garnier, O. Dédriche, M. Zekkri, M. Dorigo, Self-organized discrimination of resources. PLoS ONE **6**(5), e19888 (2010a)

A. Campo, Á. Gutiérrez, S. Nouyan, C. Pinciroli, V. Longchamp, S. Garnier, M. Dorigo, Artificial pheromone for path selection by a foraging swarm of robots. Biol. Cybern. **103**(5), 339–352 (2010b)

Y. Cao, W. Ren, Distributed coordinated tracking with reduced interaction via a variable structure approach. IEEE Trans. Autom. Control **57**(1), 33–48 (2012)

C. Castellano, S. Fortunato, V. Loreto, Statistical physics of social dynamics. Rev. Mod. Phys. **81**, 591–646 (2009)

N. Correll, A. Martinoli, Modeling and designing self-organized aggregation in a swarm of miniature robots. Int. J. Robot. Res. **30**(5), 615–626 (2011)

V. Crespi, A. Galstyan, K. Lerman, Top-down vs bottom-up methodologies in multi-agent system design. Auton. Robots **24**(3), 303–313 (2008)

A. Czirók, T. Vicsek, Collective behavior of interacting self-propelled particles. Phys. A: Stat. Mech. Appl. **281**(1–4), 17–29 (2000)

A. Czirók, A.-L. Barabási, T. Vicsek, Collective motion of self-propelled particles: kinetic phase transition in one dimension. Phys. Rev. Lett. **82**, 209–212 (1999)

H. de Vries, J.C. Biesmeijer, Self-organization in collective honeybee foraging: emergence of symmetry breaking, cross inhibition and equal harvest-rate distribution. Behav. Ecol. Sociobiol. **51**(6), 557–569 (2002)

J.-L. Deneubourg, S. Goss, Collective patterns and decision-making. Ethol. Ecol. Evol. **1**(4), 295–311 (1989)

E. Ferrante, *Information transfer in a flocking robot swarm*. Ph.D. thesis, Université libre de Bruxelles, Brussels, BE (2013)

E. Ferrante, A.E. Turgut, C. Huepe, A. Stranieri, C. Pinciroli, M. Dorigo, Self-organized flocking with a mobile robot swarm: a novel motion control method. Adapt. Behav. **20**(6), 460–477 (2012)

G. Francesca, M. Brambilla, V. Trianni, M. Dorigo, M. Birattari, Analysing an evolved robotic behaviour using a biological model of collegial decision making, ed. By T. Ziemke, C. Balkenius, J. Hallam. *From Animals to Animats 12*, vol. 7426, LNCS (Springer, 2012), pp. 381–390

G. Francesca, M. Brambilla, A. Brutschy, V. Trianni, M. Birattari, AutoMoDe: a novel approach to the automatic design of control software for robot swarms. Swarm Intell. **8**(2), 89–112 (2014)

N.R. Franks, S.C. Pratt, E.B. Mallon, N.F. Britton, D.J.T. Sumpter, Information flow, opinion polling and collective intelligence in house-hunting social insects. Philos. Trans. R. Soc. B: Biol. Sci. **357**(1427), 1567–1583 (2002)

N.R. Franks, T.O. Richardson, N. Stroeymeyt, R.W. Kirby, W.M.D. Amos, P.M. Hogan, J.A.R. Marshall, T. Schlegel, Speed-cohesion trade-offs in collective decision making in ants and the concept of precision in animal behaviour. Anim. Behav. **85**(6), 1233–1244 (2013)

R. Fujisawa, S. Dobata, K. Sugawara, F. Matsuno, Designing pheromone communication in swarm robotics: group foraging behavior mediated by chemical substance. Swarm Intell. **8**(3), 227–246 (2014)

S. Galam, Sociophysics: a review of Galam models. Int. J. Mod. Phys. C **19**(03), 409–440 (2008)

S. Garnier, F. Tache, M. Combe, A. Grimal, G. Theraulaz, Alice in pheromone land: an experimental setup for the study of ant-like robots. Proc. IEEE Swarm Intell. Symp. SIS **2007**, 37–44 (2007)

S. Garnier, C. Jost, J. Gautrais, M. Asadpour, G. Caprari, R. Jeanson, A. Grimal, G. Theraulaz, The embodiment of cockroach aggregation behavior in a group of micro-robots. Artif. Life **14**(4), 387–408 (2008)

S. Garnier, J. Gautrais, M. Asadpour, C. Jost, G. Theraulaz, Self-organized aggregation triggers collective decision making in a group of cockroach-like robots. Adapt. Behav. **17**(2), 109–133 (2009)

M. Gauci, J. Chen, W. Li, T.J. Dodd, R. Groß, Self-organized aggregation without computation. Int. J. Robot. Res. **33**(8), 1145–1161 (2014)

B.P. Gerkey, M.J. Matarić, A formal analysis and taxonomy of task allocation in multi-robot systems. Int. J. Robot. Res. **23**(9), 939–954 (2004)

S. Goss, S. Aron, J.-L. Deneubourg, J.M. Pasteels, Self-organized shortcuts in the argentine ant. Naturwissenschaften **76**(12), 579–581 (1989)

P. Grodzicki, M. Caputa, Social versus individual behaviour: a comparative approach to thermal behaviour of the honeybee (Apis mellifera L.) and the american cockroach (Periplaneta americana L.). J. Insect Physiol. **51**(3), 315–322 (2005)

Á. Gutiérrez, A. Campo, F. Santos, F. Monasterio-Huelin Maciá, M. Dorigo, Social odometry: imitation based odometry in collective robotics. Int. J. Adv. Robot. Syst. **6**(2), 129–136 (2009)

Á. Gutiérrez, A. Campo, F. Monasterio-Huelin, L. Magdalena, M. Dorigo, Collective decision-making based on social odometry. Neural Comput. Appl. **19**(6), 807–823 (2010)

H. Hamann, *Space-time continuous models of swarm robotic systems: Supporting global-to-local programming*, vol. 9, Cognitive Systems Monographs (Springer, Berlin, 2010)

H. Hamann, Towards swarm calculus: urn models of collective decisions and universal properties of swarm performance. Swarm Intell. **7**(2–3), 145–172 (2013)

H. Hamann, H. Wörn, A framework of space-time continuous models for algorithm design in swarm robotics. Swarm Intell. **2**(2–4), 209–239 (2008)

H. Hamann, T. Schmickl, H. Wörn, K. Crailsheim, Analysis of emergent symmetry breaking in collective decision making. Neural Comput. Appl. **21**(2), 207–218 (2012)

Y. Hatano, M. Mesbahi, Agreement over random networks. IEEE Trans. Autom. Control **50**(11), 1867–1872 (2005)

S. Hauert, S. Leven, M. Varga, F. Ruini, A. Cangelosi, J.C. Zufferey, D. Floreano, Reynolds flocking in reality with fixed-wing robots: communication range vs. maximum turning rate, in *Proceedings of the 2011 IEEE/RSJ International Conference on Intelligent Robots and Systems, IROS 2011*, pp. 5015–5020 (2011)

J.M. Hereford, Analysis of a new swarm search algorithm based on trophallaxis, in *2010 IEEE Congress on Evolutionary Computation, CEC* (IEEE Press, 2010), pp. 1–8

O. Holland, J. Woods, R.D. Nardi, A. Clark, Beyond swarm intelligence: the UltraSwarm, in *Proceedings of the 2005 IEEE Swarm Intelligence Symposium, SIS 2005*, pp. 217–224 (2005)

N. Jakobi, P. Husbands, I. Harvey, Noise and the reality gap: the use of simulation in evolutionary robotics, ed. By F. Morán, A. Moreno, J.J. Merelo, P. Chacón. *Proceedings of the Third European Conference on Advances in Artificial Life*, vol. 929. LNCS (Springer, 1995), pp. 704–720

R. Jeanson, S. Blanco, R. Fournier, J.-L. Deneubourg, V. Fourcassié, G. Theraulaz, A model of animal movements in a bounded space. J. Theor. Biol. **225**(4), 443–451 (2003)

R. Jeanson, C. Rivault, J.-L. Deneubourg, S. Blanco, R. Fournier, C. Jost, G. Theraulaz, Self-organized aggregation in cockroaches. Anim. Behav. **69**(1), 169–180 (2005)

S. Kernbach, R. Thenius, O. Kernbach, T. Schmickl, Re-embodiment of honeybee aggregation behavior in an artificial micro-robotic system. Adapt. Behav. **17**(3), 237–259 (2009)

H. Kitano, M. Asada, Y. Kuniyoshi, I. Noda, E. Osawa, H. Matsubara, RoboCup: a challenge problem for AI. AI Mag. **18**(1), 73–85 (1997)

J.R. Kok, N. Vlassis, Distributed decision making of robotic agents, in *Proceedings of the 8th Annual Conference of the Advanced School for Computing and Imaging*, pp. 318–325 (2003)

J.R. Kok, M.T. Spaan, N. Vlassis, Multi-robot decision making using coordination graphs, in *Proceedings of the 11th International Conference on Advanced Robotics, ICAR*, pp. 1124–1129 (2003)

S. Koos, J.-B. Mouret, S. Doncieux, The transferability approach: crossing the reality gap in evolutionary robotics. IEEE Trans. Evol. Comput. **17**(1), 122–145 (2013)

S. Kornienko, O. Kornienko, C. Constantinescu, M. Pradier, P. Levi, Cognitive micro-agents: individual and collective perception in microrobotic swarm, in *Proceedings of the IJCAI-05 Workshop on Agents in real-time and dynamic environments, Edinburgh, UK*, pp. 33–42 (2005a)

S. Kornienko, O. Kornienko, P. Levi, Minimalistic approach towards communication and perception in microrobotic swarms, in *2005 IEEE/RSJ International Conference on Intelligent Robots and Systems (IROS)*, pp. 2228–2234 (2005b)

P.J.A.M. Korst, H.H.W. Velthuis, The nature of trophallaxis in honeybees. Insectes Sociaux **29**(2), 209–221 (1982)

P.L. Krapivsky, S. Redner, Dynamics of majority rule in two-state interacting spin systems. Phys. Rev. Lett. **90**, 238701 (2003)

R. Lambiotte, J. Saramäki, V.D. Blondel, Dynamics of latent voters. Phys. Rev. E **79**, 046107 (2009)

J.A.R. Marshall, R. Bogacz, A. Dornhaus, R. Planqué, T. Kovacs, N.R. Franks, On optimal decision-making in brains and social insect colonies. J. R. Soc. Interface **6**(40), 1065–1074 (2009)

M. Massink, M. Brambilla, D. Latella, M. Dorigo, M. Birattari, On the use of Bio-PEPA for modelling and analysing collective behaviours in swarm robotics. Swarm Intell. **7**(2–3), 201–228 (2013)

M. Matarić, D. Cliff, Challenges in evolving controllers for physical robots. Robot. Auton. Syst. **19**(1), 67–83 (1996)

G. Mermoud, L. Matthey, W.C. Evans, A. Martinoli, Aggregation-mediated collective perception and action in a group of miniature robots, in *Proceedings of the 9th International Conference on Autonomous Agents and Multiagent Systems: volume 2 - Volume 2, AAMAS '10* (IFAAMAS, 2010), pp. 599–606

G. Mermoud, U. Upadhyay, W.C. Evans, A. Martinoli, Top-down vs. bottom-up model-based methodologies for distributed control: a comparative experimental study, ed. By O. Khatib, V. Kumar, G. Sukhatme.*Experimental Robotics*, vol. 79. STAR (Springer, 2014), pp. 615–629

M. Mesbahi, M. Egerstedt, *Graph Theoretic Methods in Multiagent Networks*, Princeton Series in Applied Mathematics (Princeton University Press, Princeton, 2010)

M.A. Montes de Oca, E. Ferrante, A. Scheidler, C. Pinciroli, M. Birattari, M. Dorigo, Majority-rule opinion dynamics with differential latency: a mechanism for self-organized collective decision-making. Swarm Intell. **5**, 305–327 (2011)

J. Nembrini, A. Winfield, C. Melhuish, Minimalist coherent swarming of wireless networked autonomous mobile robots, in *Proceedings of the Seventh International Conference on Simulation of Adaptive Behavior on From Animals to Animats, ICSAB* (MIT Press, 2002), pp. 373–382

S. Nolfi, D. Floreano, *Evolutionary Robotics. The Biology, Intelligence, and Technology of Self-organizing Machines* (MIT Press, Cambridge, 2000)

D. Pais, P.M. Hogan, T. Schlegel, N.R. Franks, N.E. Leonard, J.A.R. Marshall, A mechanism for value-sensitive decision-making. PLoS ONE **8**(9), 1–9 (2013)

C.A.C. Parker, H. Zhang, Cooperative decision-making in decentralized multiple-robot systems: the best-of-n problem. IEEE/ASME Trans. Mechatron. **14**(2), 240–251 (2009)

C.A.C. Parker, H. Zhang, Collective unary decision-making by decentralized multiple-robot systems applied to the task-sequencing problem. Swarm Intell. **4**, 199–220 (2010)

C.A.C. Parker, H. Zhang, Biologically inspired collective comparisons by robotic swarms. Int. J. Robot. Res. **30**(5), 524–535 (2011)

D. Payton, M. Daily, R. Estowski, M. Howard, C. Lee, Pheromone robotics. Auton. Robots **11**(3), 319–324 (2001)

D.V. Pynadath, M. Tambe, The communicative multiagent team decision problem: analyzing teamwork theories and models. J. Artif. Intell. Res. **16**(1), 389–423 (2002)

C.R. Reid, S. Garnier, M. Beekman, T. Latty, Information integration and multiattribute decision making in non-neuronal organisms. Anim. Behav. **100**, 44–50 (2015)

A. Reina, M. Dorigo, V. Trianni, Towards a cognitive design pattern for collective decision-making, ed. By M. Dorigo, M. Birattari, S. Garnier, H. Hamann, M.A. Montes de Oca, C. Solnon, T. Stützle. *Swarm Intelligence*, vol. 8667. LNCS (Springer, 2014), pp. 194–205

A. Reina, R. Miletitch, M. Dorigo, V. Trianni, A quantitative micro-macro link for collective decisions: the shortest path discovery/selection example. Swarm Intell. **9**(2–3), 75–102 (2015a)

A. Reina, G. Valentini, C. Fernández-Oto, M. Dorigo, V. Trianni, A design pattern for decentralised decision making. PLoS ONE **10**(10), e0140950 (2015b)

W. Ren, R.W. Beard, *Distributed consensus in multi-vehicle cooperative control: theory and applications*, Communications and control engineering (Springer, London, 2008)

W. Ren, R. W. Beard, E.M. Atkins, A survey of consensus problems in multi-agent coordination, in *Proceedings of the 2005 American Control Conference*, vol. 3 (IEEE Press, 2005), pp. 1859–1864

C.W. Reynolds, Flocks, herds and schools: a distributed behavioral model. ACM SIGGRAPH Comput. Graph. **21**(4), 25–34 (1987)

G. Sartoretti, M.-O. Hongler, M. de Oliveira, F. Mondada, Decentralized self-selection of swarm trajectories: from dynamical systems theory to robotic implementation. Swarm Intell. **8**(4), 329–351 (2014)

A. Scheidler, Dynamics of majority rule with differential latencies. Phys. Rev. E **83**, 031116 (2011)

A. Scheidler, A. Brutschy, E. Ferrante, M. Dorigo, The k-unanimity rule for self-organized decision-making in swarms of robots. IEEE Trans. Cybern. **46**(5), 1175–1188 (2016)

I. Schizas, G. Giannakis, S. Roumeliotis, A. Ribeiro, Consensus in ad hoc WSNs with noisy links - Part II: distributed estimation and smoothing of random signals. IEEE Trans. Signal Proc. **56**(4), 1650–1666 (2008a)

I. Schizas, A. Ribeiro, G. Giannakis, Consensus in ad hoc WSNs with noisy links - Part I: distributed estimation of deterministic signals. IEEE Trans. Signal Proc. **56**(1), 350–364 (2008b)

T. Schmickl, K. Crailsheim, Trophallaxis among swarm-robots: a biologically inspired strategy for swarm robotics. First IEEE/RAS-EMBS Int. Conf. Biomed. Robot. Biomech. BioRob **2006**, 377–382 (2006)

T. Schmickl, K. Crailsheim, Trophallaxis within a robotic swarm: bio-inspired communication among robots in a swarm. Auton. Robots **25**(1–2), 171–188 (2008)

T. Schmickl, C. Möslinger, K. Crailsheim, Collective perception in a robot swarm, ed. By E. Şahin, W.M. Spears, A.F. Winfield. *Swarm Robotics*, vol. 4433. LNCS (Springer, 2007), pp. 144–157

T. Schmickl, H. Hamann, H. Wörn, K. Crailsheim, Two different approaches to a macroscopic model of a bio-inspired robotic swarm. Robot. Auton. Syst. **57**(9), 913–921 (2009a)

T. Schmickl, R. Thenius, C. Moeslinger, G. Radspieler, S. Kernbach, M. Szymanski, K. Crailsheim, Get in touch: cooperative decision making based on robot-to-robot collisions. Auton. Agents Multi-Agent Syst. **18**(1), 133–155 (2009b)

T.D. Seeley, P.K. Visscher, T. Schlegel, P.M. Hogan, N.R. Franks, J.A.R. Marshall, Stop signals provide cross inhibition in collective decision-making by honeybee swarms. Science **335**(6064), 108–111 (2012)

O. Soysal, E. Şahin, A macroscopic model for self-organized aggregation in swarm robotic systems, ed. By E. Şahin, W.M. Spears, A.F.T. Winfield. *Swarm Robotics*, vol. 4433. LNCS (Springer, 2007), pp. 27–42

W.M. Spears, D.F. Spears, J.C. Hamann, R. Heil, Distributed, physics-based control of swarms of vehicles. Auton. Robots **17**(2), 137–162 (2004)

D.J.T. Sumpter, *Collective Animal Behavior* (Princeton University Press, Princeton, 2010)

B. Szabó, G.J. Szöllösi, B. Gönci, Z. Jurányi, D. Selmeczi, T. Vicsek, Phase transition in the collective migration of tissue cells: experiment and model. Phys. Rev. E **74**, 061908 (2006)

M. Szymanski, T. Breitling, J. Seyfried, H. Wörn, Distributed shortest-path finding by a micro-robot swarm, ed. By M. Dorigo, L.M. Gambardella, M. Birattari, A. Martinoli, R. Poli, T. Stützle. *Ant Colony Optimization and Swarm Intelligence*, vol. 4150. LNCS (Springer, 2006), pp. 404–411

V. Trianni, M. Dorigo, Emergent collective decisions in a swarm of robots, in *Proceedings of the IEEE Swarm Intelligence Symposium, SIS 2005* (IEEE Press, 2005), pp. 241–248

V. Trianni, S. Nolfi, Engineering the evolution of self-organizing behaviors in swarm robotics: a case study. Artif. Life **17**(3), 183–202 (2011)

V. Trianni, R. Groß, T.H. Labella, E. Şahin, M. Dorigo, Evolving aggregation behaviors in a swarm of robots, ed. By W. Banzhaf, J. Ziegler, T. Christaller, P. Dittrich, J.T. Kim. *Advances in Artificial Life*, vol. 2801. LNCS (Springer, 2003), pp. 865–874

V. Trianni, C. Ampatzis, A.L. Christensen, E. Tuci, M. Dorigo, S. Nolfi, From solitary to collective behaviours: decision making and cooperation, ed. By F. Almeida Costa, L.M. Rocha, E. Costa, I. Harvey, A. Coutinho. *Advances in Artificial Life*, vol. 4648. LNCS (Springer, 2007), pp. 575–584

A.E. Turgut, H. Çelikkanat, F. Gökçe, E. Şahin, Self-organized flocking in mobile robot swarms. Swarm Intell. **2**(2), 97–120 (2008)

T. Vicsek, A. Zafeiris, Collective motion. Phys. Rep. **517**(3–4), 71–140 (2012)

T. Vicsek, A. Czirók, E. Ben-Jacob, I. Cohen, O. Shochet, Novel type of phase transition in a system of self-driven particles. Phys. Rev. Lett. **75**, 1226–1229 (1995)

J. Werfel, K. Petersen, R. Nagpal, Designing collective behavior in a termite-inspired robot construction team. Science **343**(6172), 754–758 (2014)

J. Wessnitzer, C. Melhuish, Collective decision-making and behaviour transitions in distributed ad hoc wireless networks of mobile robots: target-hunting, ed. By W. Banzhaf, J. Ziegler, T. Christaller, P. Dittrich, J.T. Kim. *Advances in Artificial Life*, vol. 2801. LNCS (Springer, 2003), pp. 893–902

C.A. Yates, R. Erban, C. Escudero, I.D. Couzin, J. Buhl, I.G. Kevrekidis, P.K. Maini, D.J.T. Sumpter, Inherent noise can facilitate coherence in collective swarm motion. Proc. Nat. Acad. Sci. **106**(14), 5464–5469 (2009)

Chapter 3
Modular Design of Strategies for the Best-of-n Problem

Robot swarms frequently face discrete consensus achievement scenarios and need
to make a collective decision with the purpose to accomplish more complex tasks.
Although these applications scenarios are widespread and a consistent literature of
swarm approaches exists, they have been primarily addressed with domain-specific
methodologies. Many of them, however, share a common structure and can be cast
in the framework of the best-of-n problem. Rather than tackling these scenarios
individually, we aim at defining a general design methodology that allows us to con-
ceive collective decision-making strategies at will (e.g., with a desired compromise
of speed and accuracy). Strategies that can be transferred across different problem
domains and that have guaranteed and predictable performance. To do so, we leverage
on the idea that collective decision-making strategies can be decomposed into simple
building-blocks and we propose a modular design approach that allows the designer
to systematize the selection of each module, the derivation of predictive mathematical models, and the analysis of the swarm performance. In this chapter, we provide
a high level perspective of collective decision-making strategies and identify the
fundamental processes that characterize their functioning. Successively, we propose
a minimal modular control structure of a collective decision-making strategy and
define a set of guidelines and constraints to support the design of specific modules.
Finally, we define a general modeling methodology and show how this methodology
can be instantiated to study macroscopic properties of specific strategies.

3.1 Fundamental Mechanisms of a Strategy

At a high level of abstraction, a collective decision-making strategy is a set of control
rules that enables a swarm of agents to operate as a compact *information processing entity* with problem-solving capabilities. The swarm gathers and processes the
information available in the environment in order to make a decision on a certain
matter. When facing the best-of-n problem, each agent in the swarm has its own
preference for a certain option i, $i \in \{1, 2, \ldots, n\}$, that we call the agent's *opinion*.

© Springer International Publishing AG 2017

G. Valentini, *Achieving Consensus in Robot Swarms*, Studies in Computational
Intelligence 706, DOI 10.1007/978-3-319-53609-5_3

Throughout the entire decision-making process, the agents interact with each other and with the environment by executing the control rules prescribed by the collective decision-making strategy. As a result, the agents repeatedly reconsider and change their opinion for the best option eventually converging to a collective decision where the totality or a large majority of them share the same opinion (cf. Sect. 2.2).

The distributed information processing performed by the agents of the swarm during the collective decision-making process consists of two distinct and simultaneous processes. Campo et al. (2010) argue that:

> The decision-making process is usually a combination of exploration and information pooling that leads the group to focus its activity on one or a subset of all the available resources.

We adopt a slightly different terminology and we refer to these processes as information gathering and information pooling. At the swarm level, information gathering and information pooling are highly coupled processes simultaneously performed by different agents of the swarm.

- *Information gathering* is the process through which the agents explore the environment, discover alternative options of the decision-making problem, and individually collect information about the quality of these options.
- *Information pooling* is the process through which the agents spread throughout the swarm the information gathered about each option of the decision-making problem, sample the information of other members of the swarm and use it to reconsider their opinion.

Additional and updated information keeps flowing into the information processing entity (i.e., the swarm) during the entire decision-making process until it reaches a collective decision (see Fig. 3.1). At the agent level, information processing and information pooling are instead performed sequentially by individual agents at different times.

When executing a collective decision-making strategy, the agents interact with each other and with their environment by repeatedly applying a few control rules. A subset of these control rules, henceforth referred to as the *individual decision-making mechanism*, allows the agents to reconsider and possibly change their current opinion about the best option of the decision-making problem.

The repeated application of the individual decision-making mechanism by the agents of the swarm generates a *positive feedback* loop that leads the swarm towards

Fig. 3.1 Illustration of the swarm as an information processing entity transforming the information from the environment into a collective decision through the coupling between information gathering and information pooling

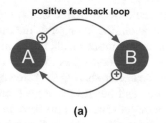
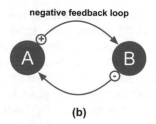

Fig. 3.2 Illustration of different feedback loops. **a** shows a positive feedback loop: an increase in the frequency of events A generates an increase of events B which further amplifies the frequency of events A and results in a snowball effect. **b** shows a negative feedback loop: the increase in the events A has the effect to increase the frequency of events B which in turns inhibits the generation of events A and stabilizes the system

one of possibly many ordered states (Garnier et al. 2007; Şahin 2005) representing different collective decisions. In general, positive feedback loops are generated by processes showing an autocatalytic behavior (Camazine et al. 2001; Deneubourg and Goss 1989). That is, from a process whose result is a catalyst event for the original process itself and determines a non-linear amplification of the resulting events (see Fig. 3.2a). Due to its autocatalytic nature, positive feedback initially builds up slowly over time by amplifying the effects of random fluctuations (e.g., a marginal majority in the opinions of the swarm favoring a certain option); then, once the process reaches a certain critical point, positive feedback increases non-linearly producing a cascade effect (e.g., a fast build-up of a majority of opinions) until experiencing a slowdown due to the depletion of the catalyst events and eventually vanishing when the system converges to an ordered state (e.g., consensus). In a collective decision-making strategy, the opinions spread by agents during information pooling and their application of the individual decision-making mechanism function as the catalysts of a self-enhanced diffusion of opinions among other members of the swarm and generates the positive feedback crucial to establish a collective decision.

In the absence of other phenomena, a swarm of agents executing a collective decision-making strategy characterized only by positive feedback loops would converge to a stable consensus decision. There are two phenomena that might prevent the swarm from reaching consensus and make it converge to a large majority. This phenomena are *negative feedback*, that results from the interactions among agents and between agents and the environment, and *intrinsic noise*, that generates internally to individual agents. Negative feedback is an inhibitory process that counterbalances the effects of positive feedback and leads the system to a homeostatic stable state (Schaber et al. 2013; Sumpter 2006). While positive feedback amplifies random fluctuations resulting in an explosive growth of the system, negative feedback has the opposite effect (see Fig. 3.2b): it stabilizes the system by inhibiting the random fluctuations that could drive the system away from a stable state (Heylighen 2001). Negative feedback usually arises from the depletion or saturation of resources available in the environment (e.g., from crowding effects that generate spatial interference and obstacle agents spatial maneuvers). When this is the case, the closer the swarm is towards a consensus decision, the stronger is the effect of negative feedback (e.g., the more robots travel along a confined path, the higher their reciprocal spatial interference). Eventually, the swarm converges towards a stable collective decision represented by a majority of the agents sharing the same opinion and corresponding to the point in which positive and negative feedback counterbalance each other. Similarly, intrinsic noise might prevent a swarm from reaching a consensus decision. Intrinsic noise generates internally within individual agents from spontaneous errors or exploratory behaviors that make an agent spontaneously change its opinion. Although both negative feedback and intrinsic noise lead to stable majority decisions, negative feedback is a non-linear process while intrinsic noise is generally linear.

Positive feedback allows a swarm to make a collective decision also when all options of the best-of-n problem have equal quality, i.e., $\rho_i = \rho_j$, $\forall i, j \in \{1, \ldots, n\}$. This result is due to the fact that positive feedback is proportional to the size of sub-populations of agents with the same opinion and amplifies random deviations from an unbiased initial condition. However, positive feedback alone is not sufficient for the swarm to make optimal collective decisions (Marshall et al. 2009) when the decision-making problem has differently valued options. To do so, the swarm needs a mechanism to process the information about the quality ρ_i of each option i and to steer the decision-making process in favor of the best option. That is, a collective decision-making strategy requires a *modulation mechanism* that acts on positive feedback in order to spread agents opinions proportionally to their quality (Garnier et al. 2007). To account for the options' quality, agents in the swarm amplify or reduce the frequency of events generating positive feedback as a function of their quality estimates (e.g., the duration of opinion dissemination, the frequency of participation to the decision-making process). As a result of the agents modulating positive feedback, the swarm has higher chances to converge on the option with highest quality as opposed to other sub-optimal alternatives.

3.2 Modular Perspective of a Strategy

We consider a modular structure of a collective decision-making strategy that implements all fundamental mechanisms necessary to solve a best-of-n problem. The primary components of the strategy structure are: *(i)* a pair of control states, respectively, the *exploration state* and the *dissemination state*; *(ii)* an individual decision-making mechanism; and *(iii)* a mechanism for the modulation of positive feedback. The exploration state implements the information gathering process and allows the agents to estimate the quality of the different options. On the other hand, the dissemination state implements the information pooling process where the gathered information is shared and processed by the agents of the swarm. These two control states are combined to form a probabilistic finite-state machine representing the basic structure of the individual agent controller. The individual decision-making mechanism and the modulation of positive feedback mechanism represent the remaining modules left to be designed. By properly selecting these two modules, the designer has a means to control the feedback processes that govern the decision-making process of the swarm.

3.2.1 Exploration and Dissemination States

The basic structure of the individual agent controller is represented by the Probabilistic Finite-State Machine (PFSM) shown in Fig. 3.3. In this agent controller, the exploration and the dissemination states are replicated for each of the n options of the decision-making problem. The resulting PFSM has $2n$ control states and, in each control state, the agents have always a preference for a certain option i of the best-of-n problem. The agents continuously alternate periods of opinion dissemination to periods of option exploration. While executing the dissemination state, the agents need to interact with each other and are therefore required to stay in a common area called *decision-making hub*. This is not required for agents executing the exploration state. An agent in the exploration state E_i explores option i for a certain period of time. After this time has elapsed, the agent transits to the dissemination state D_i (solid arrow in Fig. 3.3). At this point, the agent disseminates opinion i for a certain period of time; then, it collects the opinions disseminated by its neighbors and applies the individual decision-making mechanism to reconsider its opinion. After the individual decision, the agent transit to the exploration state corresponding to its current (possibly different) opinion (dotted arrows in Fig. 3.3).

In the exploration state E_i, the objective of an agent is to gather information from its environment—information that is instrumental in solving the decision-making problem. Specifically, an agent that favors opinion i collects a sample estimate $\hat{\rho}_i$ of the quality ρ_i (see Fig. 3.4). In case the region of the environment associated to option i does not correspond to the decision-making hub, the agent needs first to reach that region. Once it reached the region, the agent estimates the quality of

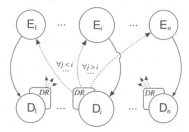

Fig. 3.3 Illustration of the individual robot control algorithm. The figure shows the PFSM for n alternative options of the decision-making problem. The transition from state E_i to state D_i is deterministic (*solid arrow*) while transitions from state D_i to states $E_j, j \in \{1, \ldots, n\}$, are stochastic (*dotted arrows*) and depend on the outcome of the individual decision-making mechanism (*DR*)

Fig. 3.4 Illustration of the control-flow followed by an agent in the exploration state E_i. Routines "Goto" and "Estimation()" are used, respectively, to move in the correct region of the environment and to sample the option quality

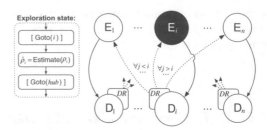

the corresponding option by means of a domain-specific routine. After collecting a sample estimate $\hat{\rho}_i$, the agent transits to the dissemination state D_i. Depending on the specific scenario, the agent might need to return to the decision-making hub before making the transition to D_i.

To successfully implement the agent's behavior in the exploration state, the designer has to define at most two routines: a relocation routine (i.e., routine "Goto()" in Fig. 3.4) and a quality estimation routine (i.e., routine "Estimate()" in Fig. 3.4). The relocation routine is not subject to particular constraints beyond that of being effective and, depending on the specific scenario, might not be required at all (i.e., when the agents can estimate the quality of all options of the best-of-n problem without leaving the decision-making hub). Contrarily, the estimation routine is always required and needs to satisfy the following condition:

- *unbiased estimation*: the collected estimate $\hat{\rho}_i$ is an unbiased, possibly noisy estimate of the option quality ρ_i, i.e., the sample mean of a collection of quality estimates $\{\hat{\rho}_i^{(1)}, \ldots, \hat{\rho}_i^{(s)}\}$ converges to the true mean ρ_i for a number s of samples that tends to infinity.

The above condition requires that the procedure followed by the agents to estimate the quality of a certain option is free from systematic errors. This condition needs to be satisfied by the particular implementation of the estimation routine. Note that the unbiased estimation condition does not require options to be associated to specific locations of the environment. All options can potentially be co-located in the same region of the environment (e.g., the decision-making hub) or be distributed in different regions.

In the dissemination state D_i, the objective of an agent is to contribute to the information pooling process performed by the swarm. Specifically, an agent with opinion i has to promote its current opinion, observe the opinions of its neighbors, and eventually leverage this information to reconsider and possibly change its opinion (see Fig. 3.5). For the entire period of opinion dissemination, the individual agent broadcasts its opinion locally (i.e., within a limited communication range) and listens to the opinions of its neighbors. At the end of the dissemination period, the agent reconsiders its opinion i about which option is the best alternative of the decision-making problem. The agent first collects the opinions promoted by its neighbors and then applies the individual decision-making mechanism as a function of the collected opinions to determine its new opinion $j \in \{1, \ldots, n\}$. Finally, the agent leaves the

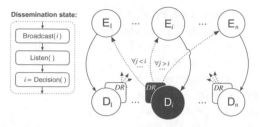

Fig. 3.5 Illustration of the control-flow followed by an agent in the dissemination state D_i. Routines "Broadcast()", "Listen()", and "Decision()" are used to disseminate opinion i, to listen to the neighbors' opinions, and to reconsider the agent's opinion

dissemination state D_i and transits to the exploration state E_j corresponding to its new opinion.

To successfully implement the agent's behavior in the dissemination state, the designer has to define three routines: an opinion dissemination routine, a routine to collect the neighbors' opinions, and a routine to reconsider the current opinion of the agent (i.e., routine "Broadcast()", routine "Listen()", and routine "Decision()" in Fig. 3.5). We focus on the necessary conditions required to implement the first two routines and postpone the discussion of the latter routine (i.e., the individual decision-making mechanism) to Sect. 3.2.3. We define three conditions:

- *decision-making hub*: the agents require a shared region of the environment where to exchange their opinions with other members of the swarm during opinion dissemination;
- *listen-only to opinion dissemination:* the agents consider only the opinion of neighboring agents that are explicitly disseminating their own opinions (i.e., explicit information transfer) and ignore instead the opinions that might be passively shared by agents in the exploration state;
- *well-mixed interaction*: within the decision-making hub, each agent has approximately the same probability to interact with any other agent in the hub independently of their opinions, i.e., the agents interaction pattern is such that the opinions are well mixed.

As already introduced at the beginning of this section, the existence of a decision-making hub ensures that agents with different opinions have the possibility to interact with each other and, in doing so, to influence each other opinion. The latter two conditions, the listen-only to opinion dissemination and the well-mixed interaction, define instead how these interactions should happen. An agent in the dissemination state collects only the information that is broadcast by other agents in the dissemination state. As will be more clear in Sect. 3.2.2, this condition is required to correctly implement the modulation of positive feedback mechanism. Additionally, the agents are required to approximate a well-mixed interaction pattern. This condition is required to prevent the spatial fragmentation of opinions within the decision-making hub (e.g., formation of clusters of robots with the same opinion i) which could prevent the system from converging to a collective decision (Deffuant et al. 2000). Depending on the specific scenario, the designer might satisfy this condition by increasing the communication range of the agents (e.g., by physically using more powerful sensors, or virtually, by making the agents function as repeaters of their neighbors' opinions) and by defining proper motion routines to stir the opinions of the agents in the decision-making hub (e.g., random walks, spatial aggregation).

On the one hand, the agent controller defined based on the dissemination and the exploration states provides the basic structure of a collective decision-making strategy. The implementation of this control structure is largely domain-specific and the designer needs only to take care of this task in order to reuse an existing strategy in different scenarios. On the other hand, the designer can separately focus on the selection of the individual decision-making mechanism and of the modulation of positive feedback mechanism. The coupling of this two components defines the performance of the collective decision-making strategy and its design is supported by macroscopic mathematical models (see Sect. 3.3). For this reason, we purposely ignored to discuss the implementation details concerning the timing of the agent controller, i.e., the duration of the exploration and dissemination states. These aspects influence the design process of the modulation of positive feedback mechanism and will be treated in Sect. 3.2.2.

3.2.2 Modulation of Feedback Loops

As introduced above, the combination of positive and negative feedback loops allows a swarm of agents to agree on a common opinion but it is the modulation mechanism that steers the collective decision in favor of the best option. Depending on the specific problem scenario, the designer needs to opportunely define a modulation mechanism that allows the agents to amplify or inhibit their individual tendency to promote a certain opinion and, in doing so, to favor the best option. As also illustrated through biological examples by Garnier et al. (2007), the factors that affect the modulation mechanism are divided into those that arise externally to the swarm, i.e., environmental bias factors and those that generate internally, i.e., internal preference factors, (cf. Sect. 2.2). We consider these factors from an engineering perspective and illustrate how they characterize the properties of the modulation mechanism and the possible choices of the designer. We distinguish between indirect and direct modulation:

- *indirect modulation:* the result of environmental bias factors influencing the behavior of the swarm, i.e., the modulation resulting from the indirect effect of the interaction between the agents and the environment with the environment being the driving force that modulates the frequency of positive and negative feedback events.
- *direct modulation:* the result of agents adjusting their behavior as a function of internal preference factors, i.e., the modulation resulting from the direct effect of

the behavior of the swarm or of its individual agents that explicitly modulate the duration of their actions generating positive and negative feedback.

Depending on the target scenario (i.e., the presence of environmental bias factors) and on the choices taken by the designer (i.e., the presence of internal preference factors), the modulation mechanism of a collective decision-making strategy can be indirect, direct or a combination of both.

The features of the environment that results in the indirect modulation of feedback loops depend on the specific scenario and are generally beyond the control of the designer. For example, in an aggregation scenario where shelters represent the alternative options of the decision-making problem, the size of each shelter influences the probability for an agent to discover it (Campo et al. 2010); in the case of the shortest-path problem, the length of each path influence the frequency with which an agent participates in the decision-making process (Montes de Oca et al. 2011; Scheidler et al. 2016); similar results would be observed with paths having equal length but different traversal times due to the asymmetric presence of obstacles or rough terrain on the paths (see Chap. 2). When environmental bias factors are present in the considered scenario, the agents are generally subject to their effects (i.e., indirect modulation) only during their execution of the exploration states E_i, $i \in \{1, \ldots, n\}$. Specifically, indirect modulation affects the duration of the exploration state. Let us consider an agent in state E_i: the shorter the time required by the agent to execute state E_i, the more frequently the agent participates to the decision-making process (and the other way around). As a consequence, opinion i is promoted more frequently and its supporters have higher chances to influence other agents in the hub whose opinion $j \neq i$. Although indirect modulation factors cannot be controlled, the designer can sometimes leverage them during the design process. When indirect modulation is negatively correlated to the quality of the different options, i.e., higher quality options have shorter exploration times, this modulation mechanism is sufficient to drive the swarm towards optimal collective decisions. The designer might choose this as a minimal implementation to simplify the collective decision-making strategy (see Chap. 4 for an example).

Differently from indirect modulation, direct modulation is the results of the actions purposely taken by individual agents owing to internal preference factors and its functioning mechanism is completely in the hand of the designer. Direct modulation is generally performed by the agents during the execution of the dissemination state: the agents adjust the duration of their opinion dissemination phase as a function positively correlated with the option quality ρ_i, i.e., higher quality options have longer dissemination times. In doing so, the agents that favor higher quality options disseminate their opinion for a longer period of time and, consequently, have higher chances to influence other members of the swarm. In Fig. 3.6, the routine "Modulate()" represents the function of ρ_i that the designer is required to define in order to implement a direct modulation mechanism. An example of direct modulation mechanism is the linear function

$$f(\hat{\rho}_i) = \hat{\rho}_i g, \tag{3.1}$$

Fig. 3.6 Illustration of the control-flow followed by an agent in the dissemination state $\mathbf{D}_i$ with direct modulation. In the implementation of the dissemination state, routine "Modulation()" determines the duration of the dissemination state and a timer mechanism ensures the determined duration

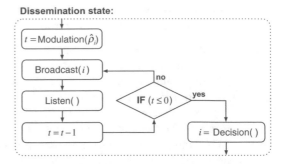

where g is a time-scaling constant (see Chaps. 5 and 6 for an extensive analysis). Alternatively, to increase the discrimination power of the modulation mechanism, the designer might select a nonlinear function, for example, the exponential function

$$f(\hat{\rho}_i) = ge^{\hat{\rho}_i k}, \tag{3.2}$$

where g and k are time-scaling constants.

As described above, indirect and direct modulation mechanisms work differently: the first modulates the duration of the time the agents spend away from the decision-making process while the second modulates the duration of the time in which they participate to it. Depending on the specific scenario, the designer might choose to couple a direct modulation mechanism to an existing indirect one to increase the performance of the collective decision-making strategy. We consider indirect modulation and distinguish two cases. In the first case, the indirect modulation mechanism is negatively correlated to ρ_i and promotes the spread of higher quality options. By coupling a direct modulation mechanism, the designer can further increase the performance of the swarm with respect to indirect modulation alone. In the second case, the indirect modulation mechanism is positively correlated to ρ_i which hinders the spread of higher quality options. This time the designer is required to use a direct modulation mechanism to reverse the situation (e.g., a linear or an exponential function of ρ_i as described in the previous paragraph). This latter scenario represents the hardest variant of the best-of-n problem from a design perspective. The performance of the direct modulation mechanism is lowered by the presence of indirect modulation and its effectiveness depends on the relative strength of the two modulation mechanisms.

3.2.3 Individual Decision-Making Mechanism

We can now focus on the last ingredient that characterizes the design of a collective decision-making strategy: the individual decision-making mechanism used by each agent of the swarm to reconsider its current opinion. From an abstract perspective, the individual decision-making mechanism is a function $h: \{1, \ldots, n\}^v \rightarrow \{1, \ldots, n\}$ that takes as input a set of opinions of size v and returns as output the new opinion of the focal agent.[1] The input of this function consists of the opinions of the neighbor agents and (possibly) that of the focal agent and needs to be collected by the focal agent prior to the application of the individual decision-making mechanism. The output of h corresponds instead to one element of the set $\{1, \ldots, n\}$ representing all possible opinions. The design of function h, that is, of the individual decision-making mechanism, strongly influences the performance of the collective decision-making strategy in terms of its speed and accuracy (Franks et al. 2003) as well as in terms of the type of the resulting collective decision (i.e., consensus versus large majority). Therefore, the designer should properly define this component in order to shape the dynamics of the collective decision-making strategy and obtain the desired performance compromise.

Depending on the choices of the designer, the individual decision-making mechanism might lead a swarm to either a stable, time-invariant consensus decision where all agents share the same opinion (Montes de Oca et al. 2011; Scheidler 2011; Scheidler et al. 2016) or to a large and fluctuating majority of agents with the same opinion but without unanimity in the swarm (Campo et al. 2010; Hamann et al. 2012; Reina et al. 2015). Consensus is favorable when the swarm of agents faces one-time or rarely-recurring decision-making problems (e.g., the selection of a construction site). Differently, a collective decision formed by a large majority of opinions allows the swarm to commit a small portion of its agents to the exploration of other options of the decision-making problem. In doing so, the swarm can detect and adapt to changes in the environment that might redefine the decision-making problem at run time (e.g., exhaustion of the available resources, discovery of a new alternative). When the objective is to obtain a stable consensus decision, the designer should satisfy the following condition:

- *consensus condition*: function h is the sole mechanism that allows an agent to change its current opinion or that of a neighbor and is such that, at all times, $h(i, \ldots, i) = i$.

The consensus condition is sufficient to obtain stable consensus decisions. It is easy to see that, once consensus is reached—let say on option i—all applications of h would always have as input a set of opinions all equal to i and would therefore return as output opinion i preventing the swarm from breaking the consensus. When the consensus condition is satisfied, the dynamics of the swarm can be described by an absorbing Markov process (see Chap. 4) whose absorbing states identify all possible collective decisions.

Popular examples of individual decision-making mechanisms that satisfy the consensus condition are the voter model, the majority rule, and the k-unanimity rule:

- *voter model*: when applying the voter model, an agent with opinion i copies the opinion j of a random agent in its neighborhood.

---

[1]Henceforth, we refer to an agent applying the individual decision-making mechanism as the focal agent.

- *majority rule*: when applying the majority rule, an agent with opinion i provided with a set of neighbors' opinions adopts the opinion j that is favored by the relative majority of the considered opinions. In case of ties between two or more options, the focal agent keeps its current opinion i.
- *k-unanimity rule*: when applying the k-unanimity rule, an agent with opinion i adopts opinion j only if all of its last k perceived neighbors had opinion j. In the case in which there is not unanimity, the agent keeps its current opinion i.

The voter model, the majority rule, and the k-unanimity rule satisfy the consensus condition and, in the absence of other phenomena, make the swarm converge to a consensus decision. However, these mechanisms differ in terms of performance and computational requirements.

The voter model is possibly the simplest mechanism for collective decision-making and has been extensively studied in the field of opinion dynamics to model processes of democratic voting and spatial conflict between different species (Clifford and Sudbury 1973; Liggett 1999). From an engineering perspective, it has low requirements since it only requires the focal agent to process one neighbor's opinion. As we will see in Chap. 5, the voter model leads to particularly accurate collective decisions but has long decision times. The majority rule is also a popular model of opinion formation extensively studied in the field of opinion dynamics (Galam 2008; Krapivsky and Redner 2003). With respect to the voter model, the majority rule requires agents to process more than one opinion; it is faster but also less accurate (see Chap. 6 for an extensive comparison of the two mechanisms). Finally, the k-unanimity rule has been introduced more recently by Scheidler et al. (2016); it has requirements similar to those of the majority rule, and its performance is a compromise between that of the voter model and that of the majority rule.

Contrarily, when the individual decision-making mechanism does not satisfy the consensus condition, the collective decision made by the swarm has the form of a large, fluctuating and possibly time-variant majority of agents sharing the same opinion i. As discussed in Sect. 3.1, this type of collective decisions might results from factors that are difficult to control by the designer (e.g., spatial interference generating negative feedback or faulty agents affected by intrinsic noise). However, the designer can purposely design an individual decision-making mechanism that violates the consensus condition to obtain a collective decision-making strategy that allows adaptation to changes of the environment. A simple approach to do so is to include the possibility of spontaneous switching in the definition of function h (see Hamann et al. 2014). In addition to a mechanism generating positive feedback (e.g., the majority rule), the agent has a certain probability to spontaneously change its current opinion i in favor of a randomly chosen option $j \in \{1, \ldots, n\}$. By properly balancing the spontaneous switching rate of individual agents, the designer can reserve a small proportion of the swarm for the exploration of the environment and the detection of its changes while the majority of the swarm exploits the currently best alternative (Hills et al. 2015). Alternatively, the designer can directly define a function h that is intrinsically stochastic as done by Garnier et al. (2009) that assigns a probability to change opinion as a function of the neighborhood size.

3.3 Modeling the Dynamics of a Strategy

In the previous section, we have defined a modular structure of a collective decision-making strategy and a set of constraints that, if satisfied, allows the designer to define a specific strategy by selecting only two modules: the modulation mechanism and the individual decision-making mechanism. Rather than proceeding in the design of these two modules by trial and error (Brambilla et al. 2013), our modular design methodology allows the designer to use a model-driven approach and to select these two modules by leveraging on the systematic derivation, analysis, and comparison of macroscopic mathematical models.

The modular structure of the agent controller introduced in Sect. 3.2.1 allows us to use a compartmental modeling approach (Godfrey 1983; Jacquez 1985) and to define a generic macroscopic model of a collective decision-making strategy. Independently of the chosen mathematical formalism (e.g., deterministic, stochastic), we consider a compartment for each possible control state of the agent controller representing the proportion (or the number) of agents of the swarm that are executing that control state. Then, we define which compartments can exchange agents with each other as defined in the PFSM of the agent controller (see Fig. 3.3). The rate (or probability) at which these exchanges of agents happen depends on the individual decision-making mechanism and on the modulation mechanism and are left as generic functions that the designer is required to specify in order to model a specific strategy. In this section, we show how this can be done by means of Ordinary Differential Equations (ODEs); nonetheless, the methodology that we propose can be adopted to define stochastic models as well (e.g., Markov Chains in Chap. 4, chemical reaction networks in Chaps. 5 and 6). Once a model for a specific collective decision-making strategy is obtained, the designer can not only analyze it to characterize the strategy's performance but she/he can use it to guide the design process by focusing on each module separately and readily comparing different design choices.

3.3.1 Generic Model Structure

Given a decision-making problem with n alternative options, we consider the proportion of agents of the swarm that are executing each of the $2n$ possible control states. We define variables e_i and d_i, with $i \in \{1, \ldots, n\}$, to model the proportion of agents with opinion i, respectively, in the exploration state E_i and in the dissemination state D_i. The dynamics resulting from the generic structure of a collective decision-making strategy defined in Sect. 3.2.1 are described by the system of $2n$ equations[2]

---

[2]It is worth noting that one of the $2n$ equations composing the system of Eq. (3.3) is redundant and could be eliminated due to the implicit constraint $\sum_{i \in \{1, \ldots, n\}} d_i + e_i = 1$ on the conservation of the swarm mass (i.e., the swarm is composed of a constant number of agents).

$$\begin{cases} \frac{d}{dt}d_i = f^i(i)e_i - f^d(\rho_i)d_i, & \forall i \in \{1, \ldots, n\}, \\ \frac{d}{dt}e_i = -f^i(i)e_i + \sum_{j=1}^n p_{ji} f^d(\rho_j)d_j, & \forall i \in \{1, \ldots, n\}. \end{cases} \tag{3.3}$$

The two equations shown above, which are repeated for each option $i \in \{1, \ldots, n\}$ of the best-of-n problem, model the evolution over time of the proportion of agents with opinion i executing, respectively, the dissemination state and the exploration state (i.e., one equation for each compartment of the model). The functions $f^d(\rho_i)$ and $f^i(i)$ model the effects of the direct and the indirect components of the modulation mechanism in terms of the rates at which different compartments exchange agents. The function $f^d(\rho_i)$ is defined by the designer and is a function of the option quality ρ_i. Contrarily, $f^i(i)$ depends on the specific scenario (cf. Sect. 3.2.2) and is not necessarily a function of ρ_i. Finally, the set of probabilities $p_{ij}, i, j \in \{1, \ldots, n\}$, models the effects of the individual decision-making mechanism and defines the probability for an agent disseminating opinion $i \in \{1, \ldots, n\}$ to change its preference in favor of opinion $j \in \{1, \ldots, n\}$ as a result of an individual decision.

Figure 3.7 illustrates the flows of agents between the different compartments of the macroscopic model defined by the system of Eq. (3.3). As a result of indirect modulation, the proportion d_i of agents in the dissemination state D_i increases with a per-agent rate $f^i(i)$ due to agents leaving the exploration state E_i (see Fig. 3.7a). The same proportion d_i decreases with a per-agent rate $f^d(\rho_i)$ as a result of direct modulation determining the timing with which agents in state D_i apply the individual decision-making mechanism and transit to any of the exploration states E_j, $j \in \{1, \ldots, n\}$. The rate with which the proportion e_i of agents in the exploration state E_i increases is the result of different contributions from each compartment d_j, $j \in \{1, \ldots, n\}$, modeling a dissemination state D_j (see Fig. 3.7b). For each of these compartments, e_i increases with a per-agent rate $p_{ji} f^d(\rho_j)$ where p_{ji} models the application of the individual decision-making mechanism by an agent with

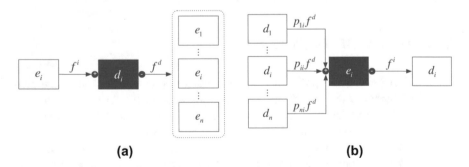

(a)   **(b)**

Fig. 3.7 Illustration of the flows of proportions of agents between different compartments of the model as defined by the system of Eq. (3.3). **a** and **b** show the flows of agents in input to and in output from the dissemination state D_i and the exploration state E_i. The *boxes* represent the proportion of agents with opinion i in the exploration state e_i, and in the dissemination state d_i, with *filled boxes* representing the focal proportion of agents. The *arrows* identify the direction and the rate of the flows of agents

opinion $j \in \{1, \ldots, n\}$. The same proportion e_i decreases with a per-agent rate $f^i(i)$ as a result of agent transitioning from the exploration state E_i to the dissemination state D_i.

3.3.2 From the Generic Structure to a Specific Model

The macroscopic model defined by the system of Eq. (3.3) is a generic model that the designer has to properly instantiate to study the dynamics of a collective decision-making strategy (i.e., a specific choice for both the modulation mechanism and the individual decision-making mechanism). In order to do so, the designer is required only to model the effects of the direct and indirect components of the modulation mechanism, i.e., functions $f^d(\rho_i)$ and $f^i(i)$, and that of the individual decision-making mechanism, i.e., probabilities $p_{ij}, i, j \in \{1, \ldots, n\}$.

The mathematical modeling of direct modulation is a relatively straightforward task for the designer. Function $f^d(\rho_i)$ defines the per-agent rate with which agents in the dissemination state D_i make individual decisions and its contribution is known a priori by the designer as a result of the design process (cf. Sect. 3.2.2). In the simplest case in which the modulation mechanism has no components implementing direct modulation, we have that the mean duration g of the dissemination state D_i is constant, independent of the option quality ρ_i, and equal for each opinion $i \in \{1, \ldots, n\}$. Therefore, function $f^d(\rho_i)$ is given by the reciprocal of g, i.e., $f^d(\rho_i) = g^{-1}$. In the general case in which direct modulation is adopted, the duration of the dissemination state is a known function of the option quality ρ_i and its reciprocal value determines $f^d(\rho_i)$. For the linear and exponential examples introduced in Sect. 3.2.2, we obtain functions

$$f^d(\rho_i) = \frac{1}{\rho_i g}, \tag{3.4}$$

and

$$f^d(\rho_i) = \frac{1}{g e^{\rho_i k}}. \tag{3.5}$$

In both equations, the parameter g is the unbiased duration of the dissemination state (set by the designer) that is modulated by the option quality ρ_i either linearly in Eq. (3.4) or exponentially in Eq. (3.5). Parameter k is a normalization constant that determines the discrimination strength of the modulation mechanism.

A similar approach is required to model the contribution $f^i(i)$ of indirect modulation which, however, is dependent on environmental features of the target scenario. This time, the designer has generally no or little a priori knowledge about the contribution of indirect modulation and is required to make assumptions based on educated guesses. In the simplest case, the designer might assume that the environment does not affect the time necessary for an agent to explore a certain option and set function $f^i(i)$ to a constant value $f^i(i) = \sigma$ independently of the agent opinion i. In the general case, the time σ_i^{-1} necessary for an agent to explore the quality of option i is different for each option $i \in \{1, \ldots, n\}$ of the best-of-n problem. Function $f^i(i)$ is therefore set to $f^i(i) = \sigma_i$. As a last example, we consider the case in which the environment positively influences the decision-making process, i.e., when the indirect modulation mechanism is a negatively correlated function of the option quality ρ_i. When this negatively correlated function is linear, as for the shortest-path problem (see Chap. 4), the contribution of indirect modulation is defined by

$$f^i(\rho_i) = \rho_i \sigma, \tag{3.6}$$

where σ is a time-scaling constant.

The next step in the design process is to model the effect of the chosen individual decision-making mechanism. In order to accomplish this task, the designer needs to find a suitable analytic description of the probability p_{ij} with which an agent with opinion i changes opinion in favor of option j by applying the individual decision-making mechanism. The definition of probability p_{ij}, $i, j \in \{1, \ldots, n\}$ for a given individual decision-making mechanism is generally a non-trivial task that requires some effort by the designer. A possible approach for the designer is to estimate the set of probabilities p_{ij} from a collection of microscopic simulations of the collective decision-making strategy. Alternatively, the designer needs to properly define an analytic description of p_{ij}. If the implementation of the agent controller satisfies the conditions defined in Sect. 3.2.1 with particular regard for the well-mixed interaction condition, then the designer has some means to define p_{ij}. In the following, we provide three examples by deriving probabilities p_{ij} for each individual decision-making mechanism introduced in Sect. 3.2.3.

In general, probability p_{ij} is a function of the current macroscopic state of the swarm, and more specifically, of the distribution of opinions among the neighbors of the agent taking an individual decision. A first step in the definition of p_{ij} is to consider the probability p_i that an agent in any of the dissemination states D_j, $j \in \{1, \ldots, n\}$, perceives the opinion i of a neighbor in state D_i. As a consequence of the well-mixed interaction condition, this probability is defined by the ratio

$$p_i = \frac{d_i}{\sum_{j=1}^{n} d_j}, \quad \forall i \in \{1, \ldots, n\}. \tag{3.7}$$

Probability p_i can be used by the designer to define the distribution of opinions among the neighbors of the focal agent and successively model the individual decision-making mechanism by properly accounting for the resulting individual decision in every different combination of opinions in the neighborhood.

As defined in Sect. 3.2.3, agents using the voter model adopt the opinion of a randomly chosen neighbor. The likelihood of this event is directly modeled by probability p_i. Similarly, when agents use the k-unanimity rule, the outcome of the individual decision depends on the opinions of k randomly chosen neighbors. The neighbors' opinions are sampled sequentially from the agents in the swarm executing the dissemination state and they trigger the focal agent to change its opinion only if they unanimously agree on the same option. If the k opinions are sampled within a short period of time, the outcome of the k-unanimity rule is well-approximated by the joint probability p_i^k (Scheidler et al. 2016). In the general case of n options, we can model the effects of the voter model and of the k-unanimity rule as

$$p_{ij} = \begin{cases} p_j^k, & \text{iff } i \neq j, \\ 1 - \sum_{\forall h, \, h \neq i} p_h^k, & \text{iff } i = j. \end{cases} \tag{3.8}$$

When parameter k is set to $k = 1$, we obtain the probabilistic description of the voter model; for values of $k > 1$, we obtain instead that of the k-unanimity rule.

Finally, when using the majority rule, agents change their opinion only in the presence of a relative majority of preferences for a certain option among their neighbors (cf. Sect. 3.2.3). We assume that, during the decision-making process, agents apply the majority rule over a group of opinions with mean size G, where G includes the opinion of the focal agent. Let us consider a focal agent with opinion i applying the majority rule over a group of opinions of mean size G. We consider the set $\mathcal{M}_{ij} = \{\langle \eta_1, \ldots, \eta_n \rangle_1, \ldots, \langle \eta_1, \ldots, \eta_n \rangle_m\}$ of all possible opinion configurations of the focal agent's neighborhood that, due to a majority decision, would result in the focal agent changing its preference from opinion i to opinion j. For each entry $\langle \eta_1, \ldots, \eta_n \rangle_h$ of the set $\mathcal{M}_{ij}$, the value of η_j, $j \in \{1, \ldots, n\}$, gives the number of agents with opinion j in that neighborhood configuration and $\sum_{j \in \{1, \ldots, n\}} \eta_j = G - 1$. The probability p_{ij} modeling the effect of the majority rule is defined as a discrete integration of a multinomial distribution

$$p_{ij} = \begin{cases} \sum_{\forall \langle \eta_1, \ldots, \eta_n \rangle \in \mathcal{M}_{ij}} \frac{(G-1)!}{\eta_1! \ldots \eta_n!} p_1^{\eta_1} \cdots p_n^{\eta_n}, & \text{iff } i \neq j, \\ 1 - \sum_{\forall h, \, h \neq i} \sum_{\forall \langle \eta_1, \ldots, \eta_n \rangle \in \mathcal{M}_{ih}} \frac{\mathcal{N}!}{\eta_1! \ldots \eta_n!} p_1^{\eta_1} \cdots p_n^{\eta_n}, & \text{iff } i = j. \end{cases} \tag{3.9}$$

Intuitively, the application of the majority rule over a certain neighborhood configuration $\langle \eta_1, \ldots, \eta_n \rangle \in \mathcal{M}_{ij}$ is modeled as sampling a number $G - 1$ of marbles (i.e., the neighborhood size excluding the focal agent opinion which is known) of a given color (opinion configuration $\langle \eta_1, \ldots, \eta_n \rangle$) from an urn (swarm of agents in the dissemination state). Due to the continuous approximation underlying the ODE model, which implies an infinite population size, sampling of each neighbor is modeled with replacement and therefore we obtain a multinomial distribution.

3.4 Discussion

In this chapter, we proposed a modular design methodology that allows a designer to define and implement a collective decision-making strategy for the best-of-n problem by focusing on the selection of few modules. Differently from trial and error approaches (Brambilla et al. 2013) or from the mimicking of specific biological behaviors (Kernbach et al. 2009; Reina et al. 2015), our design methodology leverages on the mathematical understanding of the properties of the different processes arising from the execution of a certain strategy, of which components of the strategy generate these processes, and how to control and design them. In doing so, our objective is to separately focus on the design of specific modules and obtain reusable knowledge for designing collective decision-making strategies that can be used across different problem domains. We have shown that a swarm of agents converges on a collective decision by gathering, pooling, and processing the information available in the environment, and that the resulting decision-making process is self-organized and based on the generation and modulation of positive and negative feedback loops. Along this process, we identified and formalized the basic structure of a collective decision making strategy, i.e., a combination of exploration and dissemination states, and its two fundamental modules, the modulation mechanism and the individual decision-making mechanism, that determine the strategy performance.

Our design methodology relies on the definition of a generic agent controller that the designer is required to implement for each specific problem scenario. We defined the agent controller as a PFSM that makes the agent continuously alternate between a period of option exploration and a period of opinion dissemination. This agent controller provides the basic structure of a collective decision-making strategy and implements the information gathering and information pooling necessary to the swarm to make a collective decision. If the implementation of the designer satisfies a set of constraints, the remaining part of the design process is represented by the selection of the modulation mechanism and of the individual decision-making mechanism. We have illustrated how the modulation of positive and negative feedback loops influences the outcome of the collective decision of the swarm. We characterized the properties of a modulation mechanism and distinguished between direct modulation, where agents directly amplify or inhibit the feedback loops, and indirect modulation, where the modulation is instead a result of the features of the environment that affect the behavior of the swarm. Similarly, we showed that the fact that the individual decision-making mechanism leads to a collective decision corresponding to consensus or to a large majority of opinions depends both on the choices of the designer and on the features of the environment.

This modular perspective of a collective decision-making strategy allows the designer to develop a large part of the design process with a model-driven approach that is independent of the target scenario. The basic structure of the agent controller allows us to define a generic macroscopic model of a collective decision-making strategy by adopting a compartmental modeling approach and systematically derive equations for each state of the agent controller. Finally, we provided several examples of how the designer can instantiate the macroscopic model for a specific choice of individual decision-making mechanism and modulation mechanism. In the next part of this book, we illustrate our modular model-driven design methodology by extensively analyzing and comparing different design alternatives and by showing how the designed collective decision-making strategies can be implemented in different swarm robotics scenarios.

References

M. Brambilla, E. Ferrante, M. Birattari, M. Dorigo, Swarm robotics: a review from the swarm engineering perspective. Swarm Intell. **7**(1), 1–41 (2013)

S. Camazine, J.-L. Deneubourg, N.R. Franks, J. Sneyd, G. Theraulaz, E. Bonabeau, *Self-Organization in Biological Systems* (Princeton University Press, Princeton, 2001)

A. Campo, S. Garnier, O. Dédriche, M. Zekkri, M. Dorigo, Self-organized discrimination of resources. PLoS One **6**(5), e19888 (2010)

P. Clifford, A. Sudbury, A model for spatial conflict. Biometrika **60**(3), 581–588 (1973)

G. Deffuant, D. Neau, F. Amblard, G. Weisbuch, Mixing beliefs among interacting agents. Adv. Complex Syst. **3**(01n04), 87–98 (2000)

J.-L. Deneubourg, S. Goss, Collective patterns and decision-making. Ethol. Ecol. Evol. **1**(4), 295–311 (1989)

N.R. Franks, A. Dornhaus, J.P. Fitzsimmons, M. Stevens, Speed versus accuracy in collective decision making. Proc. R. Soc. B: Biol. Sci. **270**, 2457–2463 (2003)

S. Galam, Sociophysics: a review of Galam models. Int. J. Mod. Phys. C **19**(03), 409–440 (2008)

S. Garnier, J. Gautrais, G. Theraulaz, The biological principles of swarm intelligence. Swarm Intell. **1**(1), 3–31 (2007)

S. Garnier, J. Gautrais, M. Asadpour, C. Jost, G. Theraulaz, Self-organized aggregation triggers collective decision making in a group of cockroach-like robots. Adapt. Behav. **17**(2), 109–133 (2009)

K. Godfrey, *Compartmental Models and Their Application* (Academic Press, Cambridge, 1983)

H. Hamann, T. Schmickl, H. Wörn, K. Crailsheim, Analysis of emergent symmetry breaking in collective decision making. Neural Comput. Appl. **21**(2), 207–218 (2012)

H. Hamann, G. Valentini, Y. Khaluf, M. Dorigo, Derivation of a micro-macro link for collective decision-making systems: uncover network features based on drift measurements, in *Parallel Problem Solving from Nature – PPSN XIII*, vol. 8672, LNCS, ed. by T. Bartz-Beielstein, J. Branke, B. Filipič, J. Smith (Springer, Berlin, 2014), pp. 181–190

F. Heylighen, The science of self-organization and adaptivity, *Knowledge Management, Organizational Intelligence and Learning, and Complexity of the Encyclopedia of Life Support Systems* (EOLSS Publishers Co. Ltd., Oxford, 2001), pp. 253–280

T.T. Hills, P.M. Todd, D. Lazer, A.D. Redish, I.D. Couzin, Cognitive Search Research Group, Exploration versus exploitation in space, mind, and society. Trends Cogn. Sci. **19**(1), 46–54 (2015)

J.A. Jacquez, *Compartmental Analysis in Biology and Medicine* (University of Michigan Press, Ann Arbor, 1985)

S. Kernbach, R. Thenius, O. Kernbach, T. Schmickl, Re-embodiment of honeybee aggregation behavior in an artificial micro-robotic system. Adapt. Behav. **17**(3), 237–259 (2009)

P.L. Krapivsky, S. Redner, Dynamics of majority rule in two-state interacting spin systems. Phys. Rev. Lett. **90**, 238701 (2003)

T.M. Liggett, Stochastic interacting systems: contact, voter and exclusion processes, *Grundlehren der mathematischen Wissenschaften*, vol. 324 (Springer, Berlin, 1999)

J.A.R. Marshall, R. Bogacz, A. Dornhaus, R. Planqué, T. Kovacs, N.R. Franks, On optimal decision-making in brains and social insect colonies. J. R. Soc. Interface **6**(40), 1065–1074 (2009)

M.A. Montes de Oca, E. Ferrante, A. Scheidler, C. Pinciroli, M. Birattari, M. Dorigo, Majority-rule opinion dynamics with differential latency: a mechanism for self-organized collective decision-making. Swarm Intell. **5**, 305–327 (2011)

A. Reina, G. Valentini, C. Fernández-Oto, M. Dorigo, V. Trianni, A design pattern for decentralised decision making. PLoS One **10**(10), e0140950 (2015)

E. Şahin, Swarm robotics: from sources of inspiration to domains of application, in *Swarm Robotics*, vol. 3342, LNCS, ed. by E. Şahin, W. Spears (Springer, Berlin, 2005), pp. 10–20

J. Schaber, A. Lapytsko, D. Flockerzi, Nested autoinhibitory feedbacks alter the resistance of homeostatic adaptive biochemical networks. J. R. Soc. Interface **11**(91), 20130971 (2013)

A. Scheidler, Dynamics of majority rule with differential latencies. Phys. Rev. E **83**, 031116 (2011)

A. Scheidler, A. Brutschy, E. Ferrante, M. Dorigo, The k-unanimity rule for self-organized decision-making in swarms of robots. IEEE Trans. Cybern. **46**(5), 1175–1188 (2016)

D.J.T. Sumpter, The principles of collective animal behaviour. Philos. Trans. R. Soc. B: Biol. Sci. **361**(1465), 5–22 (2006)

Part II
Mathematical Modeling and Analysis

Chapter 4
Indirect Modulation of Majority-Based Decisions

We investigate an example of collective decision-making strategy that entirely relies on the environment for the modulation of positive feedback. We consider a strategy originally proposed in the context of collective transport and show how it can be reinterpreted as *Indirect Modulation of Majority-based Decisions*. In this collective decision-making strategy, the majority rule is coupled to the spatial asymmetries of the environment. Our aim is to study this strategy with an approach able to accurately capture the dynamics of finite swarms regardless of their size. We use the formalism of time homogeneous Markov chains that allows us to consider swarms of any finite size and to estimate the accuracy and the speed of the collective decision.

4.1 Problem Scenario and Decision-Making Strategy

The majority rule with differential latency model has been originally developed in the context of collective transport in robot swarms (Montes de Oca et al. 2011) and applied to a binary decision-making problem that resembles the well-known double-bridge experiment performed by Goss et al. (1989). Robots in the swarm need to collectively decide between two possible actions to perform, henceforth referred to as action a and action b. Actions differ in their execution times and the goal of the swarm is to reach consensus on the action with the shortest execution time. Specifically, robots in the swarm need to transport objects from a *source* area to a *destination* area (see Fig. 4.1). To this end, robots can choose between two possible paths: path a and path b. Choosing and traversing a path corresponds to performing action a or action b. The two paths differ in length and therefore are characterized by different traversal times (i.e., the path length is the environmental bias factor, cf. Sect. 2.2). Additionally, each object is too heavy for a single robot to be transported

© Springer International Publishing AG 2017
G. Valentini, *Achieving Consensus in Robot Swarms*, Studies in Computational Intelligence 706, DOI 10.1007/978-3-319-53609-5_4

Fig. 4.1 Illustration of the collective transport scenario. Robots can travel back and forth between the source and the destination areas by traversing either path *a* (*top*) or path *b* (*bottom*)

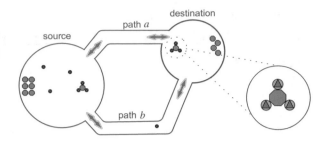

and requires instead a team of 3 robots. Once a team if formed, the robots collectively decide which path to take. In the following, we summarize the functioning of this strategy and reinterpret it according to the framework introduced in Chap. 3.

4.1.1 Control Algorithm

We consider a swarm composed of N robots all executing the same control algorithm. Each robot in the swarm has an opinion that defines its currently favored path. During the execution of the collective transport behavior, a robot can change its opinion as a result of the interaction with other members of the swarm. At any given time, a robot is either inactive, waiting for other robots in the source area, or is actively engaged in the transport process and is traversing a path. Montes de Oca et al. (2011) refer to the robots that are waiting in the source area as *non-latent* robots and to those that are traversing a path as *latent* robots. Only non-latent robots can take part in the decision-making process and influence other members of the swarm. In order to transport objects from the source area to the destination area, non-latent robots form teams of 3 members each. Once a team is formed, robots decide which path to traverse. To do so, robots share with the other members of the team their opinions concerning their favored path; then, each robot applies the majority rule over this set of opinions and determines its newly favored path (and therefore that of the entire team[1]). At this stage, the robots become latent, i.e., they leave the source area as a team transporting an object, and travel back and forth along the chosen path. On its return from a journey, the team disbands and its members become non-latent joining other robots resting in the source area.

The majority rule with differential latency model has been designed by taking inspiration from opinion formation models (Galam 1986; Krapivsky and Redner 2003) rather than by following the modular design approach introduced in Chap. 3. Specifically, Montes de Oca et al. (2011) extended the idea of latency introduced by Lambiotte et al. (2009) in the canonical majority rule model. In the model of Lambiotte et al. (2009), when an agent switches opinion as a consequence of

---

[1]Note that, due to the team's size of 3 robots, the application of the majority rule always results in a team's consensus due to the impossibility of decision ties.

the application of the majority rule it turns into a latent state for a latency period that has a stochastic duration. Later, Montes de Oca et al. (2011) extended this idea by using different latency periods as a function of the option quality and, in doing so, they obtain the discrimination capabilities necessary to select the best option of the best-of-n problem. Despite the different sources of inspiration, we can identify in the majority rule with differential latency model the same structure of a collective decision-making strategy previously introduced in Chap. 3. Specifically, we can reinterpret this strategy as an *Indirect Modulation of Majority-based Decisions* (IMMD).

The distinction of agents between those that are latent and those that are non-latent corresponds to our definition of the exploration state and the dissemination state. Robots that are non-latent and resting in the source area corresponds to agents disseminating their opinions in the decision-making hub. In this strategy, the duration of the dissemination state is independent of the opinions of individual agents. Therefore, agents are not provided with a mechanism for the direct modulation of positive feedback. The majority rule is adopted as the individual decision-making mechanism and its application by triplet of agents after the formation of a team marks the agents' transition from the dissemination state to the exploration state. When robots are latent, i.e., during their trip back and forth between the source and destination areas, they are executing the exploration state aimed at sampling the quality of the available options. The process of estimating the quality of a certain option of the decision-making problem is not performed by agents directly measuring the length or the traversal time of a particular path (cf. Montes de Oca et al. 2011). Instead, the spatial asymmetries naturally present in the environment (i.e., the length of a path) indirectly modulate the positive feedback which steers the decision-making process towards the best option. In this system, the shorter the time spent by an agent to traverse a path (i.e., to explore the corresponding option), the more frequently the agents will return to the source area. As a consequence, agents that favor the opinion associated to the best available option have higher chances to influence other members of the swarm with their opinions, in this way positively affecting the decision-making process.

As in Montes de Oca et al. (2011), we consider a scenario where the number of latent robots is constant and a multiple k of the team size 3. Additionally, we consider the duration of the latency period to be exponentially distributed and its expected value to be a function of the opinion (i.e., the path) favored by the team of robots. In our analysis, we make use of the terminology defined in Chap. 3 in order to provide the reader with a homogeneous discussion of the topic. In particular, we refer to the latent and non-latent states as the exploration and dissemination states. Additionally, we refer to the expected values of the latency periods, which are the reciprocal of the options' qualities ρ_a and ρ_b, as the expected duration of the exploration state, respectively, for path a and for path b.

4.1.2   Monte Carlo Simulation

In order to validate the results of our study, we implemented a simple Monte Carlo simulation of the IMMD strategy. We simulated two sets of agents: agents in the exploration state grouped into teams of 3 members and characterized by an opinion and a duration of the exploration state proportional to the option quality (i.e., ρ_a or ρ_b); and agents in the dissemination state described only by their opinions. The Monte Carlo simulation proceeds at discrete time steps; at each time step the following instructions are executed:

1. the exploring team having minimum residual duration of the exploration state is disbanded, its composing agents are added to the set of agents in the dissemination state, and its residual exploration time is subtracted to that of all other teams in the exploration state;
2. 3 agents are randomly sampled without replacement from the set of agents in the dissemination state and form a team;
3. the opinions of the sampled agents are noted and the majority rule is applied over this set to determine the new opinion i of the team;
4. the new team is added to the set of agents in the exploration state and its exploration time is randomly drawn from an exponential distribution with expected value determined by $1/\rho_i$.

The simulation schema defined above is repeated until the swarm reaches consensus over any of the two options. In order to validate the predictions of the Markov chain model, we average the results of Monte Carlo simulations over 1000 independent repetitions for each combination of the parameters of the system.

4.2   Markov Chain Model

We model the execution of the IMMD strategy by defining an absorbing, time-homogeneous Markov chain[2] with a finite state space (Kemeny and Snell 1976). Similarly to (Montes de Oca et al. 2011), we assume a constant number k of teams executing the exploration state. Additionally, without loss of generality, we consider the expected duration of the exploration periods to have mean values $1/\rho_a = 1$, for opinion a, and $1/\rho_b$, $0 \leqslant \rho_b \leqslant 1$, for opinion b. That is, opinion a is either the best option of the best-of-2 problem or both options are equally-good (i.e., when $\rho_b = 1$). We consider each application of the individual decision-making mechanism as one step of the decision-making process along the chain. More precisely, we consider each step ϑ as being composed of three stages:

1. A team from those in the exploration state transits to the dissemination state (i.e., it finishes its exploration period).

---

[2]The reader may refer to Appendix A for a minimal background on time-homogeneous Markov chains.

2. A new team of 3 agents is randomly formed out of the set of agents in the dissemination state.
3. The agents in the team apply the majority rule to determine the team's opinion. Next, the agents transit as a team to the exploration state.

Given a particular choice of values for the swarm size N and for the number k of teams in the exploration state, we aim at studying the evolution of the number of agents with opinion a over ϑ. Let $\mathbb{N}$ represent the set of natural numbers. We define the state of the Markov chain as a tuple $s = \langle E_a; D_a \rangle$, where $E_a \in \{E_a : E_a \in \mathbb{N}, 0 \leqslant E_a \leqslant k\}$ is the number of teams in the exploration state with opinion a and $D_a \in \{D_a : D_a \in \mathbb{N}, 0 \leqslant D_a \leqslant N - 3k\}$ is the number of agents in the dissemination state with opinion a. The number E_b of teams in the exploration state with opinion b and the number D_b of agents in the dissemination state with opinion b is obtained by difference, respectively as $k - E_a$ and $N - 3k - D_a$. Note that each state of the Markov chain provides a macroscopic perspective of the opinions within the swarm. The resulting state space of the Markov chain is characterized by $m = (k + 1)(N - 3k + 1)$ states (i.e., the number of elements in the Cartesian product of the domains of E_a and D_a). In the following, we use symbols s_i and s_j to refer to a pair of generic states of the chain, while we use symbols s_a and s_b to refer to the consensus states $\langle k; N - 3k \rangle$ and $\langle 0; 0 \rangle$ in which the entire swarm agrees on opinion a and b, respectively. Note that s_a and s_b are absorbing states of the Markov chain; consequently, once the decision-making process reaches one of these states it will remain trapped therein for all future times (cf. Kemeny and Snell 1976). In practice, the absorption of the Markov process into one of the absorbing states s_a and s_b identifies the achievement of consensus and the completion of the decision-making process.

At the generic step ϑ, the decision-making process moves from state $s_i = \langle E_a^{(i)}; D_a^{(i)} \rangle$ to state $s_j = \langle F_a^{(j)}; D_a^{(j)} \rangle$ following the aforementioned 3 stages. At stage 1, a team from those in the exploration state finishes its exploration period, transits to the dissemination state and disbands. The probability p_i that this team has opinion a is given by

$$p_i = \frac{\rho_a E_a^{(i)}}{\rho_a E_a^{(i)} + \rho_b (k - E_a^{(i)})} = \frac{E_a^{(i)}}{E_a^{(i)} + \rho_b (k - E_a^{(i)})}. \tag{4.1}$$

The set of agents in the dissemination state with opinion a increases of $c = 3$ units, if the disbanding team has opinion a and of $c = 0$, if the disbanding team has opinion b. At stage 2, a new team is formed by 3 random agents from the set of agents in the dissemination state (i.e., those agents resting in the source area). We are interested in the probability q_i that the new team is formed by a number $0 \leqslant g \leqslant 3$ of agents with opinion a. As in Sect. 3.3.2, the probability q_i is defined by a hypergeometric distribution,

$$q_i(g; c) = \frac{\binom{d_i + c}{g} \binom{N - 3k - D_a^{(i)} + 3 - c}{3 - g}}{\binom{N - 3k + 3}{3}}. \tag{4.2}$$

In the above equation, $N - 3k + 3$ represents the number of opinions in the (current) set of agents in the dissemination state which also include the 3 agents of the most recently disbanded team. This set of opinions is composed of $D_a^{(i)} + c$ preferences for opinion a and $N - 3k - D_a^{(i)} + 3 - c$ preferences for opinion b. Finally, at stage 3, the majority rule is applied by the robots in the newly formed team and its outcome is represented by the value of g. Eventually, the decision-making process moves to the next state s_j.

Equations (4.1) and (4.2) allow us to define the transition probabilities between each possible pair of states s_i and s_j. These probabilities are the entries of the stochastic transition matrix P, which completely defines the dynamics of a Markov process (cf. Kemeny and Snell 1976). However, not all pairs of states identify a feasible step of the process along the chain according to the rules of the system, i.e., not all pair of states are *adjacent* to each other. Two states s_i and s_j are adjacent if $\Delta_{ij}s = \langle \Delta_{ij}E_a = E_a^{(j)} - E_a^{(i)}; \Delta_{ij}D_a = D_a^{(j)} - D_a^{(i)} \rangle$ appears in the first column of the following table. The correspondent transition probability P_{ij} is given in the second column:

$\langle \Delta_{ij}E_a; \Delta_{ij}D_a \rangle$	P_{ij}	stage 1	stage 2
$\langle -1; 3 \rangle$	$p_i q_i (3 - \Delta_{ij}d; 3)$	a	$3b$
$\langle -1; 2 \rangle$		a	$a2b$
$\langle 0; 1 \rangle$		a	$2ab$
$\langle 0; 0 \rangle$	$p_i q_i (3 - \Delta_{ij}d; 3) + (1 - p_i) q_i (|\Delta_{ij}d|; 0)$	a	$3a$
		b	$3b$
$\langle 0; -1 \rangle$	$(1 - p_i) q_i (|\Delta_{ij}d|; 0)$	b	$a2b$
$\langle 1 : -2 \rangle$		b	$2ab$
$\langle 1; -3 \rangle$		b	$3a$

Columns three and four provide the corresponding events observed in stage 1 and stage 2, i.e., the opinion of the agents in the next team returning from the exploration of an option and the opinions of the agents that randomly form a new team in the set of agents in the dissemination state. For values of $\Delta_{ij}s$ not included in column one, the transition probability is $P_{ij} = 0$.

The probabilistic interpretation of P is straightforward: at any step ϑ, if the decision-making process is in state $s(\vartheta) = s_i$ it will move to state $s(\vartheta + 1) = s_j$ with probability P_{ij}. It is worth noticing that, being the consensus states s_a and s_b two absorbing states, the probability mass of s_a and s_b is concentrated in the corresponding diagonal entries of P, that is, $P_{aa} = 1$ and $P_{bb} = 1$.

4.3   Analysis of Opinion Dynamics

In order to analyze the dynamics of the majority rule with differential latency, we follow Kemeny and Snell (1976) and define, on the basis of the transition matrix P, the matrices Q, R, and F. Matrix Q describes transitions between pairs of transient states, matrix R gives the probability to move from a transient state to an absorbing state, and matrix $F = (I - Q)^{-1}$ is the *fundamental* matrix with matrix I being the identity matrix. From matrices Q, R, and F of the Markov chain model we study the dynamics of a finite swarm. We validate the predictions of our model with the results of Monte Carlo simulations averaged over 1000 independent repetitions for each combination of the parameters of the system. The interested reader may refer to Appendix A.2 for further details on the derivation of Q, R, and F as well as other mathematical derivations used in this section.

4.3.1 Exit Probability

We are interested in studying the accuracy of the IMMD strategy in terms of its capability to discriminate the best option of the best-of-2 problem. To reach this goal, we derive the exit probability E_N, i.e., the probability that a swarm of N agents that starts the execution of the IMMD strategy with the initial configuration $s(\vartheta = 0) = s_i$ reaches consensus on the opinion associated to the shortest path—opinion a. This probability is given by the entries associated to the consensus state s_a of the matrix of the absorption probabilities resulting from the multiplication FR.

Figure 4.2 shows the predictions of the exit probability over the initial proportion $(3E_a(0) + D_a(0))/N$ of agents favoring opinion a for different configurations of the parameters of the system. A first observation that can be drawn from the results shown in Fig. 4.2a is that, for increasing values of the swarm size N, the exit probability E_N approaches a step function. This step function is centered around the critical density that divides the initial configurations of the swarm leading to a decision for

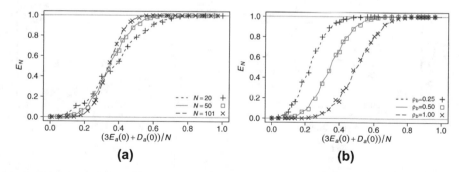

Fig. 4.2 Exit probability E_N that the swarm reaches consensus on opinion a against the initial proportion $(3E_a(0) + D_a(0))/N$ of agents favoring that opinion for different parameter configurations of the system. **a** Show the value of the exit probability for a decision-making problem characterized by $\rho_b = 0.5$ and swarm size $N \in \{20, 50, 101\}$. **b** Shows the exit probability when varying the problem difficulty $\rho_b \in \{0.25, 0.5, 1.0\}$ for a swarm of size $N = 50$. The lines provide the predictions of the Markov chain model while the symbols correspond to the results of 1000 Monte Carlo simulations for each initial configuration of the system

option a (i.e., the best option) from those leading to a decision for option b (i.e., the worst option). Additionally, as also found by Scheidler (2011), we can observe from Fig. 4.2b that the larger is the expected duration $1/\rho_b$ of the exploration state associated to opinion b, the smaller is the initial number of preferences for opinion a that results in the exit probability being biased towards option a (i.e., $E_N \geqslant 0.5$). We observe a good accuracy in the predictions of the Markov chain model (lines) when compared to the average behavior of the Monte Carlo simulations (symbols) regardless of the number of agents in the swarm.

4.3.2 Expectation and Variance of Consensus Time

In addition to the exit probability, we are interested in characterizing the time necessary for a swarm to make a collective decision. This quantity is related to the number τ of applications of the majority rule that the agents in the swarm have to take in order to reach consensus. As specified in Sect. 4.2, each step of the decision-making process along the Markov chain corresponds to one application of the individual decision-making mechanism—the majority rule. In the following, we use our Markov chain model to compute the expected value of τ as $\hat{\tau} = \xi F$, where ξ is a column vector of all 1s (cf. Appendix A.2). The entries of $\hat{\tau}$ correspond to the row sums of the fundamental matrix F. In turn, F gives the mean sojourn time for each transient state of a Markov chain, that is, the expected number of times that a process started in state $s(\vartheta = 0) = s_i$ transits through state s_j. The variance of τ is defined as $\hat{\tau}_2 = (2F - I)\hat{\tau} - \hat{\tau}_{sq}$, where the matrix I is the identity matrix and the

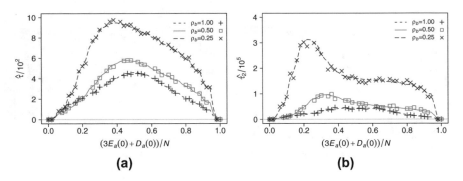

Fig. 4.3 Distribution of the number of applications of the majority rule necessary to reach consensus against the initial proportion $(3E_a(0) + D_a(0))/N$ of agents favoring opinion a for a swarm of $N = 50$ agents and $k = 16$ teams when the decision-making problem is characterized by $\rho_b \in \{1, 0.5, 0.25\}$. **a** Shows the expected value $\hat{\tau}$ of the number of applications of the majority rule; **b** shows instead its variance $\hat{\tau}_2$. The lines provide the predictions of the Markov chain model while the symbols correspond to the average of 1000 Monte Carlo simulations for each initial configuration of the system

vector $\hat{\tau}_{sq}$ corresponds to $\hat{\tau}$ with squared entries. Additional details can be found in Appendix A.2.

Figure 4.3a, b show, respectively, the expected value $\hat{\tau}$ and the variance $\hat{\tau}_2$ of the number of applications of the majority rule necessary to reach consensus for a swarm of $N = 50$ agents. As for the exit probability analyzed in the previous section, the Markov chain model predicts the Monte Carlo simulations with good accuracy. The expected value $\hat{\tau}$ of the number of applications of the majority rule reaches its maximum near the critical density that divides initial configurations of the systems that lead to consensus on opinion a from those that lead to consensus on opinion b. The results shown in Fig. 4.3a highlight a critical feature of the IMMD strategy: in order to make a collective decision, the swarm requires a number of applications of the majority rule that is larger for easier discrimination problems (i.e., $\rho_b \rightarrow 0$) than for more difficult ones (i.e., $\rho_b \rightarrow 1$). This result is due to the fact that agents traveling along the sub-optimal path, i.e., path b, need to travel for a longer time (than agents traveling along path a) before returning to the decision-making hub and have a chance to change their opinions; meanwhile, agents with opinion a keep applying the majority rule which affects the value of $\hat{\tau}$.

The predictions of our Markov chain are in agreement with the analysis of consensus time developed in Scheidler (2011). However, we can observe from Fig. 4.3b that the expected number $\hat{\tau}$ of applications of the majority rule (which is related to the consensus time studied by Scheidler (2011)) does not provide a faithful description of this system. Indeed, the variance $\hat{\tau}_2$ of the number of application of the majority rule is about three orders of magnitude larger than the expected value $\hat{\tau}$. We note that this behavior is characteristic of absorbing Markov processes (cf. Kemeny and Snell 1976). Similarly to the expected value, we observe that easier decision-making problems are characterized by a larger variance of the number of applications of the majority rule.

4.3.3 Distribution of Consensus Time

Finally, we derive the cumulative distribution function $P(\tau \leqslant \vartheta; s_i)$ of the number of decisions necessary to the swarm to reach consensus as well as its probability mass function $P(\tau = \vartheta; s_i)$. From a swarm robotics perspective, we are interested in the dynamics of a system initially unbiased, i.e., a swarm that begins the execution of the decision-making process with an equal number of preferences for both the opinion a and the opinion b. Let $s(\vartheta = 0) = s_u$ represents this initial unbiased configuration. Recalling that matrix Q is the matrix of the transition probabilities for the transient states, we have that the entries of Q_{uj}^{ϑ}, i.e., the ϑth power of matrix Q, give the probabilities that the decision-making process is in the transient state s_j at step ϑ when started in state s_u. Thus, the row sum of the u-th row of Q^{ϑ} gives the probability that the decision-making process is still in one of the transient states at step ϑ. From this probability, we can derive the cumulative distribution function $P(\tau \leqslant \vartheta; s_u)$ by computing the series $\{1 - \sum_j Q_{uj}^{\vartheta}\}$ for values of ϑ such that $Q^{\vartheta} \rightarrow 0$.

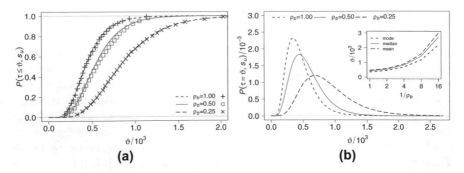

Fig. 4.4 Distribution of the number of applications of the majority rule necessary to reach consensus for a swarm of $N = 50$ agents and $k = 16$ teams when the decision-making problem is characterized by $\rho_b \in \{1, 0.5, 0.25\}$. **a** Shows the cumulative distribution function $P(\tau \leqslant \vartheta; s_u)$ of the number of applications of the majority rule; **b** Shows instead the probability mass function $P(\tau = \vartheta; s_u)$ with details of the mode, the median and the mean values. The lines provide the predictions of the Markov chain model while the symbols correspond to the average of 1000 Monte Carlo simulations for each initial configuration of the system

Figure 4.4a shows the cumulative distribution function $P(\tau \leqslant \vartheta; s_u)$ for a swarm of $N = 50$ agents that begins the decision-making process unbiased. In agreement with the results shown in Fig. 4.3, the longer is the expected duration $1/\rho_b$ of the exploration state associated to opinion b (i.e., the lower is the option quality ρ_b), the larger is the number of applications of the majority rule necessary to reach consensus. Figure 4.4b, provides the probability mass function $P(\tau = \vartheta; s_u)$, together with details of the mode, the median, and the mean values of τ. As shown in the inset of Fig. 4.4b, the values of the mode, the median and the mean statistics diverge for increasing values of the ratio $\rho_a/\rho_b = 1/\rho_b$ between the expected duration of the exploration state associated to opinion a and that associated to opinion b. Moreover, when $1/\rho_b \to \infty$ the shape of the distribution $P(\tau = \vartheta; s_u)$ tends to a flat function revealing therefore that the variance dominates the system.

4.4 Discussion

In this chapter, we considered the majority rule with differential latency (Montes de Oca et al. 2011) and provided an equivalent interpretation of this strategy according to the modular design methodology introduced in Chap. 3. We referred to this strategy as *Indirect Modulation of Majority-based Decisions* (IMMD) to highlight its constituent modules: an indirect mechanism for the modulation of positive feedback and the majority rule.

The majority rule with differential latency model (i.e., the IMMD strategy) is a benchmark model of collective decision-making in robot swarms and has been the subject of a large number of research studies (Montes de Oca et al. 2011;

Scheidler 2011; Massink et al. 2012, 2013). Most of these studies resulted in the definition of mathematical models that provide continuous approximations of the swarm dynamics (e.g., deterministic ODE models derived in the limit of an infinite swarm size (Montes de Oca et al. 2011) and continuous approximations of finite size systems Scheidler 2011). The primary limitation of continuous approximations is that the defined mathematical models provide accurate and reliable predictions only when the number of agents in the swarm is relatively large. Conversely, swarm robotics aims to design scalable control strategies that operate for swarms of any size, ranging from tens to millions of agents. Moreover, it is usually difficult to derive statistics from a continuous approximation model that go beyond the expected value of quantities of interest and this often results in poorly informative descriptions of the underlying swarm dynamics.

We designed an absorbing Markov chain model to predict the performance of the IMMD strategy in a swarm composed of a finite number of agents. Using our model, we studied the probability that a system of N agents reaches consensus on the opinion associated to the best option of the decision-making problem (i.e., the shortest path), as well as the distribution of the number of applications of the majority rule necessary to reach consensus. This latter result reveals that this collective decision-making strategy is characterized by a large variance of the number of decisions necessary before consensus, and thus, that its expected value, which was used as a measure of performance in previous studies, is a relatively poor statistic for this system. In contrast to continuous approximations, we explicitly model the macroscopic state space of the system—which is discrete—and the transition probabilities governing its dynamics. This approach allows us to derive reliable predictions of the swarm dynamics regardless of its size.

Our contribution is relevant from a swarm robotics perspective because it allows us to advance the understanding of the IMMD strategy with respect to previous studies. The analysis of our Markov chain model, with particular regard to the distribution of the number of applications of the majority rule necessary to reach consensus, provides the possibility to perform statistical inference on certain interesting aspects of the system. For example, using the results in Fig. 4.4, we can compute the probability that the swarm has reached consensus as a function of time and use this information to improve the efficiency of quorum-detection mechanisms (Parker and Zhang 2010). Moreover, the approach can be easily extended to model individual decision-making mechanisms other than the majority rule, allowing for a priori comparison of different design choices.

The IMMD strategy studied in this chapter provides an example of collective decision-making strategy for discrete consensus achievement problems that couples the majority rule (which functions as the individual decision-making mechanism) with a mechanism for the indirect modulation of positive feedback (which results from an environmental bias). Different compromises between the speed and the accuracy of a collective decision could be achieved by employing individual decision-making mechanisms other than the majority rule (e.g., the k-unanimity rule discussed in Chap. 3). However, the primary limitation of this collective decision-making strategy results from the fact that it is entirely dependent on environmental bias factors for the modulation of positive feedback and these factors are beyond the control of the designer. As a consequence, its applicability is limited to those scenarios in which the environment naturally defines the quality of alternative options and in which the best option is the one characterized by the shortest exploration time. In the next chapter, we will explore an example of collective decision-making strategy that, in contrast to the IMMD strategy, makes use of a mechanism for the direct modulation of positive feedback.

References

S. Galam, Majority rule, hierarchical structures, and democratic totalitarianism: a statistical approach. J. Math. Psychol. **30**(4), 426–434 (1986)

S. Goss, S. Aron, J.-L. Deneubourg, J.M. Pasteels, Self-organized shortcuts in the argentine ant. Naturwissenschaften **76**(12), 579–581 (1989)

J.G. Kemeny, J.L. Snell, *Finite Markov Chains* (Springer, New York, 1976)

P.L. Krapivsky, S. Redner, Dynamics of majority rule in two-state interacting spin systems. Phys. Rev. Lett. **90**, 238701 (2003)

R. Lambiotte, J. Saramäki, V.D. Blondel, Dynamics of latent voters. Phys. Rev. E **79**, 046107 (2009)

M. Massink, M. Brambilla, D. Latella, M. Dorigo, M. Birattari, Analysing robot decision-making with Bio-PEPA, in *Swarm Intelligence*, LNCS, vol. 7461 ed. by M. Dorigo, M. Birattari, C. Blum, A.L. Christensen, A.P. Engelbrecht, R. Groß, T. Stützle (Springer, New York, 2012), pp. 25–36

M. Massink, M. Brambilla, D. Latella, M. Dorigo, M. Birattari, On the use of Bio-PEPA for modelling and analysing collective behaviours in swarm robotics. Swarm Intell. **7**(2–3), 201–228 (2013)

M.A. Montes de Oca, E. Ferrante, A. Scheidler, C. Pinciroli, M. Birattari, M. Dorigo, Majority-rule opinion dynamics with differential latency: a mechanism for self-organized collective decision-making. Swarm Intell. **5**, 305–327 (2011)

C.A.C. Parker, H. Zhang, Collective unary decision-making by decentralized multiple-robot systems applied to the task-sequencing problem. Swarm Intell. **4**, 199–220 (2010)

A. Scheidler, Dynamics of majority rule with differential latencies. Phys. Rev. E **83**, 031116 (2011)

Chapter 5
Direct Modulation of Voter-Based Decisions

We study a collective decision-making strategy where the modulation of positive feedback results directly from the actions of individual agents of the swarm. We consider a binary site-selection scenario and focus on the study of the *Direct Modulation of Voter-based Decisions*. This collective decision-making strategy combines the voter model with the direct modulation of positive feedback. The voter model is a model of opinion dynamics extensively studied in the field of statistical physics. In the classic voter model agents change opinion by adopting the opinion of a random neighbor. We couple the voter model with a mechanism for the direct modulation of positive feedback that is inspired by the waggle dance behavior of honeybees. We study this strategy by means of an ordinary differential equation model, a chemical reaction network, and multi-agent simulations. This set of macroscopic and microscopic models enables us to investigate the behavior of the swarm both in the thermodynamic limit of an infinite swarm size and when random fluctuations arise as a consequence of a finite swarm size. Based on our results, we provide the requirements on the initial conditions necessary to guarantee a consensus decision on the best option of the decision-making problem as well as the minimum swarm size necessary to guarantee a given level of accuracy. Finally, we show that this strategy scales with the size of the swarm and is robust in the presence of noise in the agent estimates of the option quality.

5.1 Problem Scenario and Decision-Making Strategy

In the following, we focus on the study of binary decision-making problems (i.e., the best-of-n problem with $n = 2$) and we refer to the two alternative options as option a and option b. Specifically, we consider a site-selection scenario (see Fig. 5.1) where the goal of the swarm is to select the best site available in the environment where to relocate the swarm. We consider agents acting within a bounded, two-

© Springer International Publishing AG 2017

G. Valentini, *Achieving Consensus in Robot Swarms*, Studies in Computational Intelligence 706, DOI 10.1007/978-3-319-53609-5_5

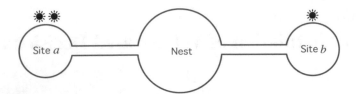

Fig. 5.1 Illustration of the site selection scenario. Robots can travel back and forth between the nest (at the *center*) and the two sites, respectively, site *a* (on the *left*) and site *b* (on the *right*). In this example, the quality of a site consists in the level of brightness of the ambient light and is represented by a sun symbol over the site, that is, site *a* is twice as good as site *b*

dimensional environment which is divided in a number of regions. Each option of the decision-making problem is associated to a particular region in the environment that is referred to as *site*. The opinion of each agent in the swarm represents its preference for a particular site in the environment. In addition to sites, the environment is characterized by a third region called the *nest*. The nest functions as a hub for the decision-making process and all agents are initially located in this region of the environment. Agents repeatedly travel between the nest and their currently favored site. The time necessary to travel towards and back from a site is the same for both sites. As a consequence, the decision-making problem has no environmental bias factor that might influence the modulation of positive feedback. Once in a certain site, an agent can perceive and estimate the quality of that site (i.e., an internal preference factor). Without loss of generality, we consider site *a* to have higher quality than site *b* and for the remaining of this chapter we set $\rho_a = 1$ and we vary the value of ρ_b in the interval (0, 1]. We consider the decision-making problem successfully solved if the swarm reaches consensus on opinion *a*—the opinion associated to the best site.

5.1.1 Control Algorithm

In the *Direct Modulation of Voter-based Decisions* (DMVD), agents in the swarm are driven by the probabilistic finite-state machine (PFSM) shown in Fig. 5.2 (i.e., a specific instance of the generic PFSM shown in Fig. 3.3 of Chap. 3). In the *dissemination* states (either D_a or D_b) agents advertise their currently favored site by locally broadcasting their opinion and eventually apply the individual decision-making mechanism, i.e., the voter model, to reconsider their opinion. In the *exploration* states (either E_a or E_b) agents estimate the quality of the site associated to their current opinion. The time spent by an agent in a certain state consists of two contributions. Initially, agents spend an unknown period of time necessary to reach the proper region of the environment where to perform the activities defined by their current state (henceforth referred to as *traveling time*). Next, once arrived in the correct region, agents perform the actions associated to their current state for a period of time defined by a parameter set by the designer. We choose to adopt exponentially distributed periods of time. Due to its memory-less property, the exponential distribution eases our successive

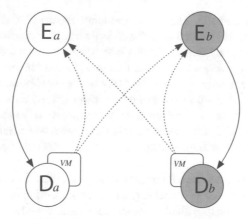

Fig. 5.2 Illustration of the probabilistic finite-state machine of the individual agent. *Solid* and *dotted lines* represent, respectively, deterministic and stochastic transitions; symbols D_i and E_i with $i \in \{a, b\}$ represent the dissemination states and the exploration states, while symbols *VM* highlight the application of the voter model at the end of the dissemination state

derivations of mathematical models and enhances the predictability of the proposed strategy. Alternative options to this choice could be provided by constant time periods or by stochastic time periods whose probability distribution is different from the exponential distribution. However, these latter alternatives are characterized by less favorable mathematical properties.

As soon as agents transit to the dissemination state (i.e., either D_a or D_b) they begin to perform a random walk within the boundaries of the nest. In the meanwhile, agents advertise their opinion about their currently favored site. Agents purposely perform a random walk in order to mix their opinions within the nest and prevent spatial fragmentation of opinions that might hinder the decision-making process. Before their transition to the exploration state, that is, as soon as the dissemination time expires, agents reconsider their opinion about the best available site. As in the classic voter model (Clifford and Sudbury 1973; Liggett 1999), an agent polls the opinions of its neighbors in the dissemination state within a limited interaction range; then, the agent adopts the opinion of a randomly picked neighbor from this poll. Similarly to honeybees (cf. Visscher and Camazine 1999), the individual decision-making mechanism used by the agents does not take into consideration the quality of the advertised sites. Once agents have deliberated on their opinion, they transit to the exploration state (i.e., either E_a or E_b, see Fig. 5.2).

A core mechanism of the DMVD strategy, which implements the selection of the best option, is the modulation of positive feedback (Garnier et al. 2007). The agent controller is designed to scale the time spent in the dissemination states proportionally to the quality of the opinions. The time spent disseminating opinion a (respectively, opinion b) is directly proportional to the opinion's quality $\rho_a g$ (respectively, $\rho_b g$) where g is the unbiased dissemination time, a parameter set by the designer. The parameter g represents the average duration of opinion dissemination without considering its modulation; the designer needs to choose the value of g depending on the specific application scenario. The agents modulate the positive feedback by setting the amount of time during which they disseminate a certain opinion to be an increasing function of the option quality. In this way agents influence the frequency with which other agents observe a certain opinion in their dissemination state. As a consequence, observing neighbors that are in favor of the best option is more likely than observing neighbors that are in favor of other, lower quality alternatives. Therefore, the swarm is biased towards achieving a collective decision for the best option. This idea is loosely inspired by the honeybee behavior shown when honeybees search for potential site locations for their new nest (Franks et al. 2002; Seeley 2010; von Frisch 1967).

As soon as agents enter the exploration state (i.e., either E_a or E_b) they leave the nest and move toward the site associated with their current opinion. Once arrived at the correct site, the agent evaluates the characteristic features that determine the quality associated to its opinion following a given domain-specific routine. The quality-estimation routine depends on the particular target scenario and could involve complex agent behaviors—for example, those necessary to explore a candidate construction site and evaluate its level of safety. Independently of the scenario, the quality estimation routine results in one sample measurement which is generally subject to noise. The swarm processes noisy measurements by acting as a filter that averages over many individual agent measurements. Once the exploration is completed, the agent switches to the dissemination state that corresponds to its current opinion (cf. solid lines in Fig. 5.2).

As introduced in Sect. 3.2.1, a requirement of the DMVD strategy is that the interaction among agents in the dissemination state should be well-mixed or, at least, approximately well-mixed. That is, the probability of any agent to encounter a neighbor of a certain opinion is proportional to the distribution of opinions in the whole swarm. The well-mixed property is only a weak requirement as it influences the efficiency of the decision-making process but only in extreme cases its efficacy. If the spatial distribution of agents is sufficiently well-mixed, the decision-making strategy is efficient and successful. The more the system deviates from a well-mixed state, the slower the decision-making process is. Only if the spatial distribution of agents is far from well-mixed, then the decision-making process is slowed down considerably by spatial fragmentation of opinions (e.g., formation of clusters of agents with the same opinion) and might even end up in a deadlock, that is, a macroscopic state of indecision far from consensus (Deffuant et al. 2000). In Chap. 7, we will explain how this requirement can be fulfilled for the case of autonomous ground-based robots.

5.1.2 Multi-agent Simulation

We implemented a simple multi-agent simulator with the aim to study the dynamics of a swarm using the DMVD strategy. In our simulations, agents are represented as mass-less particles, i.e., they are points moving at constant velocity in a bounded,

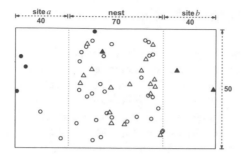

Fig. 5.3 Illustration of a multi-agent simulation of the site-selection scenario. The environment is divided in three different regions: site a on the *left* side, the nest in the *center*, and site b on the *right* side. Agents of the swarm are represented by *colored* symbols: o for agents in state D_a, • for agents in state E_a, △ for agents in state D_b, and ▲ for agents in state E_b

two-dimensional space. As a consequence, we do not consider a particular metric system or scale for the size of the environment but we employ dimensionless space units.

Agents are positioned in a rectangular arena of 150×50 space units (see Fig. 5.3). The arena is divided into three regions: two regions of 40×50 space units at the two extremities of the arena represent the sites, respectively, site a on the left side and site b on the right side; and a region of 70×50 space units centered between the two sites represents the nest. Agents are equipped with a digital compass that, when necessary, allows them to reorient their motion toward a particular region of the environment. In Fig. 5.3, we represent agents' opinions by colored symbols using red circles for opinion a and blue triangles for opinion b. Empty symbols represent agents in the dissemination state (either D_a or D_b) and filled symbols represent agents in the exploration state (either E_a or E_b).

In our simulations, agents perform the control algorithm described in Sect. 5.1.1. Their motion is determined by a random walk implemented as follows. Agents move straight for a normally distributed period of time with mean duration of 2 s and a standard deviation of 0.33 s; next, they change their direction of motion by uniformly choosing a new random orientation in the interval $[-\pi; \pi]$ and then resume straight motion. Since we model agents as dimensionless points, we do not consider collisions between agents in our simulations. However, agents do collide with the boundaries of the arena. In the case of a collision with a wall, the agent changes its direction of motion by mirroring the angle of incidence.

Finally, when applying the individual decision-making mechanism, agents first pool the opinions of neighboring agents in the dissemination state (i.e., only those that are advertising their opinion) within a given interaction range r. Then, they randomly choose one of the opinions within their pool. In the case that an agent has no neighbors, thus being unable to survey other opinions, the agent keeps its current opinion. After applying the individual decision-making mechanism, agents transit to the exploration state associated to their current opinion (i.e., either E_a or E_b).

5.2 Ordinary Differential Equations Model

In the thermodynamic limit, i.e., when the number N of agents tends to infinity, $N \to \infty$, random fluctuations that characterize the behavior of self-organizing systems vanish and the system itself approaches a deterministic behavior. Such an asymptotic perspective allows us to gain insights into the dynamics of a swarm executing the DMVD strategy regardless of the swarm size. Our object of interest is the development of consensus and therefore we look at the dynamics of the opinions in space (i.e., in the nest and in the sites) and in time. We make use of dynamical systems theory and we define a system of ordinary differential equations (ODEs) with the aim to describe the dynamics of the DMVD strategy. In our derivations, we assume null traveling times and we consider only the contributions of the design parameters[1] g and σ^{-1} (cf. Chap. 3).

We define quantities d_a and d_b as proportions of agents in the swarm that are in the dissemination state advertising, respectively, site a and site b (i.e., they have respectively opinion a and opinion b). Additionally, we denote the proportion of agents exploring site a using the symbol e_a and the proportion of agents exploring site b using the symbol e_b. The evolution over time of the proportions of agents d_a, d_b, e_a and e_b is given by

$$
\begin{cases}
\dfrac{d}{dt}d_a = \sigma e_a - \alpha d_a, \\[2mm]
\dfrac{d}{dt}d_b = \sigma e_b - \beta d_b, \\[2mm]
\dfrac{d}{dt}e_a = p_{aa}\alpha d_a + p_{ba}\beta d_b - \sigma e_a, \\[2mm]
\dfrac{d}{dt}e_b = p_{ab}\alpha d_a + p_{bb}\beta d_b - \sigma e_b.
\end{cases}
\tag{5.1}
$$

In Eq. (5.1), which is a specialization of Eq. (3.3) provided in Chap. 3, symbols $\alpha = (\rho_a g)^{-1}$ and $\beta = (\rho_b g)^{-1}$ model the modulation of the dissemination time by the option quality, respectively, for opinion a and for opinion b; probabilities p_{ij}, $i, j \in \{a, b\}$, model the application of the voter model. Given the probability $p_a = d_a/(d_a + d_b)$ that an agent adopts opinion a by randomly choosing the opinion of a neighboring agent, we have that $p_{aa} = p_{ba} = p_a$ and $p_{ab} = p_{bb} = 1 - p_a$, cf. Sect 3.3. We recall that probability p_a is derived under the assumption of a well-mixed population of agents in the nest. That is, we assume that each agent has the same probability to interact with every other agent in the nest.

The flows of the proportions of agents between the different control states can be understood by looking at Fig. 5.4. The proportion d_a of agents advertising site a (respectively, d_b for site b) increases at a rate σe_a due to the agents returning from the exploration of site a (respectively, site b). This proportion also decreases at a

---

[1] We recall that parameter g corresponds to the unbiased dissemination time before modulation and parameter σ^{-1} corresponds to the mean exploration time.

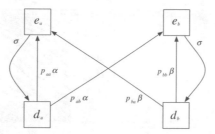

Fig. 5.4 Illustration of the flows of proportions of agents between pairs of control states. The proportion of agents in the different control state are represented by symbols d_a and d_b, respectively for the dissemination states, and by symbols e_a and e_b, respectively for the exploration states. The labels of the *arrows* give the rates of the flows between pairs of control states

rate αd_a as agents leave the dissemination state associated to opinion a (respectively, at a rate βd_b for d_b). The dynamics of the proportion e_a of agents exploring site a (the reasoning is equivalent for site b) depends on the application of the individual decision-making mechanism of the DMVD strategy. Therefore, it depends on probabilities p_{ij}. Specifically, the proportion e_a increases at a rate $p_{aa}\alpha d_a + p_{ba}\beta d_b$ due to the agents leaving the dissemination state after adopting opinion a; proportion e_a decreases at a rate σ due to the agents that have completed their exploration of site a.

In the ODE model, the magnitude and the ratio of control parameters g and σ determine the duration of the collective decision process. The longer the time agents spend at the sites or at the nest the longer is the consensus time. At design time, the designer should carefully set the value of the dissemination time g in order to approximate a well-mixed distribution of agents in the nest. Once the designer has chosen a value for the dissemination time g, the consensus time of the DMVD strategy increases linearly with the value of σ^{-1}. From an engineering perspective of minimizing the consensus time, a designer should, when possible, favor minimal values for the control parameters g and σ^{-1} such that $g \gg \sigma^{-1}$. In Fig. 5.5a, we compare the predictions of the ODE model (lines) with the results of 10^3 independent multi-agent simulations for a swarm of $N = 10^3$ agents and infinite interaction range $r = \infty$ (box-plots). We set the quality of the sites to $\rho_a = 1.0$ and $\rho_b = 0.875$. The difference between opinion qualities drives the system toward consensus on the best opinion ($d_a + e_a = 1$ and $d_b + e_b = 0$). The results in Fig. 5.5a show a good agreement between the predictions of the ODE model and the behavior of the multi-agent simulations.

In the thermodynamic limit, the DMVD strategy guarantees consensus on the best option of the decision-making problem. Figure 5.5b shows the evolution over time of the proportion $d_a + e_a$ of agents in the swarm with opinion a for a number of different initial conditions. When $\rho_a > \rho_b$, every trajectory initially starting at $\{d_a \in (0, 1], d_b = 1 - d_a\}$ eventually converges to consensus on opinion a (that is, $d_a + e_a = 1$). As it happens in the classic voter model (Clifford and Sudbury 1973; Liggett 1999), the two macroscopic solutions, consensus either on opinion a or on opinion b, characterize the asymptotic behavior of the collective decision-making

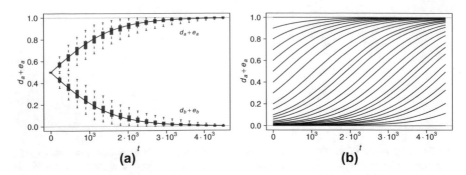

Fig. 5.5 Illustration of the transitory behavior of a swarm of agents executing the DMVD strategy as predicted by the ODE model. **a** shows the time evolution of the proportion $d_a + e_a$ of agents with opinion a predicted by the ODE model (*lines*) compared with the results of multi-agent simulations (*box-plots*). **b** shows the time evolution of $d_a + e_a$ for a number of different initial conditions. Parameters: $N = 1000$, $r = \infty$, $g = 100\,\mathrm{s}$, $\sigma^{-1} = 10\,\mathrm{s}$, $\rho_a = 1$, and $\rho_b = 0.875$

process. Notably, for the assumption $\rho_a = 1$ and $\rho_a > \rho_b$, the consensus $d_a + e_a = 1$ is a stable fixed point, while the consensus $d_a + e_a = 0$ is an unstable fixed point. The system of ODEs is characterized by the two fixed points

$$\gamma_1^\star = \left[d_a = \frac{g}{g + \sigma^{-1}}, d_b = 0, e_a = \frac{\sigma^{-1}}{g + \sigma^{-1}}, e_b = 0 \right], \tag{5.2}$$

$$\gamma_2^\star = \left[d_a = 0, d_b = \frac{\rho_b g}{\rho_b g + \sigma^{-1}}, e_a = 0, e_b = \frac{\sigma^{-1}}{\rho_b g + \sigma^{-1}} \right]. \tag{5.3}$$

The fixed point $\gamma_1^\star$ is an asymptotically stable point and the system converges to it for all initial conditions $d_a \in (0, 1]$, $d_b = 1 - d_a$. The fixed point $\gamma_2^\star$ is instead an unstable point and the swarm would reach $\gamma_2^\star$ only if initially started with consensus on opinion b ($d_a = 0$, $d_b = 1$). The analytic formulations of $\gamma_1^\star$ and $\gamma_2^\star$ provided in Eqs. (5.2) and (5.3) might be useful in real-world applications because they allow to tune at design time the final distribution of agents between the nest and the best site (e.g., to optimize the collection of resources in foraging tasks Montes de Oca et al. 2011; Scheidler 2011).

Finally, we consider the non-equilibrium dynamics of a swarm of agents executing the DMVD strategy by looking at the speed of change $d/dt\,(d_a + e_a)$ of the proportion of agents with opinion a as a function of itself ($d_a + e_a$). Figure 5.6 depicts a number of trajectories for different values of the ratio σ^{-1}/g of control parameters and various initial conditions $\{d_a \in (0, 1], d_b = 1 - d_a, e_a = 0, e_b = 0\}$ (shaded lines). Initially, the value of ($d_a + e_a$; $d/dt\,d_a + d/dt\,e_a$) is determined by the initial conditions of the system of ODE and is independent of the ratio σ^{-1}/g (see cross symbols in Fig. 5.6, each trajectory moves from the left to the right). The speed of change of the proportion of agents with opinion a as a function of itself decreases abruptly due to agents rapidly redistributing among the nest and the two sites. At a later stage of the

Fig. 5.6 Illustration of the speed of change of the proportion $d_a + e_a$ of agents favoring opinion a as predicted by the ODE model. Parameters: $g = 100$ s, $\sigma^{-1} \in \{10, 50, 150\}$ s, $\rho_a = 1$, and $\rho_b = 0.875$

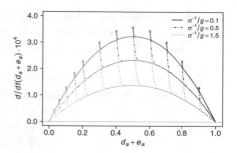

decision-making process, trajectories converge towards a parabolic curve determined by the magnitude of σ^{-1}/g. The less time agents spend to explore the sites, the faster is the change in the proportion of agents with opinion a (compare solid and dotted lines). Note that the speed of change of the proportion of agents with opinion a reaches its peak at the unbiased conditions $d_a + e_a = 0.5$. The distribution of opinions among the agents in the swarm deviates rapidly from the unbiased scenario with $d_a + e_a = 0.5$ and converges towards consensus $d_a + e_a = 1$ with decreasing speed.

5.3 Chemical Reaction Network Model

In Sect. 5.2, we studied the properties of the DMVD strategy in the thermodynamic limit of $N \to \infty$ where the swarm approaches a deterministic behavior. Real-world swarm systems, however, are composed of a large but finite number of agents. In many of these systems, finite size crucially influences the system's dynamics so that predictions based on continuous approximations might be of limited use (Toral and Tessone 2007). A number of different modeling techniques exist to deal with finite-size effects, such as Markov chains (Hamann 2013; Soysal and Şahin 2007; Valentini and Hamann 2015) and master equations (Lerman et al. 2005; Martinoli et al. 1999, 2004; Scheidler et al. 2016; Vigelius et al. 2014). In this chapter, we use the formalism of chemical master equations which are derived from a chemical reaction network (van Kampen 1992). Note that the predictions of the chemical reaction network defined below converge to those of the ODE model for increasing values of the swarm size N. As a consequence, the two models can be considered equally good approximations of the collective decision-making process. However, the chemical reaction network has the advantage of providing a greater descriptive power than the ODE model and is more accurate for small swarm sizes N.

Chemical master equations are stochastic differential equations modeling the dynamics of coupled chemical reactions among a set of molecules. Using this formalism to model a multi-agent system, agents in different states are represented by molecules of different types, while state transitions of individual agents are represented by chemical reactions with certain rates. Chemical master equations are often hard if not impossible to solve analytically. For this reason, we base our study on numerical simulations using the Gillespie algorithm (Gillespie 1977). The Gillespie algorithm—also known as Stochastic Simulation Algorithm—generates statistically correct trajectories of a master equation which can be used to approximate its exact solution.

Given a swarm of N agents, we use symbol $\mathfrak{D}_i, i \in \{a, b\}$, to represent an agent in one of the two dissemination states and symbol $\mathfrak{E}_i, i \in \{a, b\}$, to represent an agent in one of the two exploration states. Symbols $\mathfrak{D}_a$, $\mathfrak{D}_b$, $\mathfrak{E}_a$ and $\mathfrak{E}_b$ are the equivalent of the molecule species generally used in the definition of chemical reactions (Gillespie 1977). Similarly, we define the number $D_i, i \in \{a, b\}$, of agents in the nest that are disseminating their preferences for site i and the number $E_i, i \in \{a, b\}$, of agents exploring site i. The DMVD strategy is described by a set of chemical reactions and corresponding reaction rates. In the following, we provide the equations for the reactions concerning agents that favor opinion a (reactions modeling agents with opinion b are derived similarly). Within the nest, agents change opinion as a result of the application of the voter model. Such a change in the opinion is captured by reactions

---

Algorithm 5.1: Gillespie algorithm for the DMVD strategy. Parameters: $(N, D_a(0), \sigma, \alpha, \beta)$.

---

1 Initialize agents in the swarm: $D_a = D_a(0)$, $D_b = N - D_a(0)$, $E_a = 0$, $E_b = 0$
2 Initialize time: $t = 0$
3 **repeat**
4   Compute the total reaction rate: $\kappa = \alpha D_a + \beta D_b + \sigma E_a + \sigma E_b$
5   Generate an exponentially distributed time t' with rate parameter κ and set $t = t + t'$
6   Set $p_{aa} = \frac{D_a}{D_a + D_b}$ and $p_{ba} = \frac{D_a}{D_a + D_b}$
7   Let x be a uniformly distributed random number in the range $[0, 1)$
8   **if** $x \in [0; p_{aa}\alpha D_a/\kappa)$ $\longrightarrow$ Set $D_a = D_a - 1$ and $E_a = E_a + 1$
9   **if** $x \in [p_{aa}\alpha D_a/\kappa; \alpha D_a/\kappa)$ $\longrightarrow$ Set $D_a = D_a - 1$ and $E_b = E_b + 1$
10   **if** $x \in [\alpha D_a/\kappa, \alpha D_a/\kappa + p_{ba}\beta D_b/\kappa)$ $\longrightarrow$ Set $D_b = D_b - 1$ and $E_a = E_a + 1$
11   **if** $x \in [\alpha D_a/\kappa + p_{ba}\beta D_b/\kappa; \alpha D_a/\kappa + \beta D_b/\kappa)$ $\longrightarrow$ Set $D_b = D_b - 1$ and $E_b = E_b + 1$
12   **if** $x \in [\alpha D_a/\kappa + \beta D_b/\kappa; \alpha D_a/\kappa + \beta D_b/\kappa + \sigma E_a/\kappa)$ $\longrightarrow$ Set $E_a = E_a - 1$ and $D_a = D_a + 1$
13   **if** $x \in [\alpha D_a/\kappa + \beta D_b/\kappa + \sigma E_a/\kappa; 1)$ $\longrightarrow$ Set $E_b = E_b - 1$ and $D_b = D_b + 1$
14 **until** $D_a + E_a \in \{0, N\}$;

$$\mathfrak{D}_a \xrightarrow{\alpha} \mathfrak{E}_A | \mathfrak{E}_b \iff \begin{cases} \mathfrak{D}_a \xrightarrow{p_{aa}\alpha} \mathfrak{E}_a \\ \mathfrak{D}_a \xrightarrow{p_{ab}\alpha} \mathfrak{E}_b. \end{cases} \tag{5.4}$$

Probabilities p_{aa} and p_{ab} (respectively, p_{ba} and p_{bb}) are defined as in Sect. 5.2 by means of the probability $p_a = D_a/(D_a + D_b)$ that an agent adopts opinion a by copying the opinion of a random neighbor. As defined in Sect. 5.1.1, agents with opinion a reconsider their current opinion at a rate of $\alpha = (\rho_a g)^{-1}$. When doing so, agents keep opinion a with probability $p_{aa} = p_a$ and switch to opinion b with probability $p_{ab} = 1 - p_a$. Agents ceasing to explore site a are modeled by the chemical reaction

$$\mathfrak{E}_a \xrightarrow{\sigma} \mathfrak{D}_a. \tag{5.5}$$

Equation (5.5) models the agents leaving the exploration state E_a and moving to the dissemination state D_a with a constant rate σ.

The set of chemical reactions defined by Eqs. (5.4) and (5.5), together with the respective chemical reactions for agents with opinion b, suffice to define a master equation following the methods described by van Kampen (1992). The solution of the master equation can be studied numerically by applying the Gillespie algorithm—a Markov chain Monte Carlo method capable of generating statistically correct trajectories of stochastic equation (Gillespie 1977). Algorithm 5.1 depicts our particular formulation of the Gillespie algorithm that models the behavior of a finite swarm of agents executing the DMVD strategy.

In the remaining of this section, we analyze the master equation by means of Algorithm 5.1 and we assess how finite-size effects influence the dynamics of the DMVD strategy. For this purpose, we consider (*i*) the exit probability E_N, i.e., the probability that a swarm of N agents eventually reaches consensus over opinion a and (*ii*) the consensus time T_N, i.e., the time necessary to reach consensus on any opinion. The exit probability E_N is computed as the fraction of trajectories that converge to consensus on opinion a over the overall number of executions of Algorithm 5.1; the consensus time T_N is computed as the average time necessary to reach consensus over all executions of the algorithm. In our analysis of the DMVD strategy we set the unbiased dissemination time to $g = 100\,\mathrm{s}$ and the average exploration time to $\sigma^{-1} = 10\,\mathrm{s}$ and we vary the value of the parameters N, ρ_b and r. The numerical solutions of the master equation model are compared against the results of agent-based simulations both averaged over $2.5 \cdot 10^4$ independent executions.

5.3.1 Influence of Option Qualities

The difficulty of the site-selection scenario is a function of the quality of the various options of the decision-making problem. Both the accuracy of the collective decision and the time necessary to reach this decision depend on the option qualities (ρ_a and ρ_b). In this section, we study this relationship by analyzing the chemical reaction network defined above for values of $\rho_a = 1.0$ and $\rho_b \in (0, 1]$.

We consider a swarm of $N = 100$ agents with unlimited interaction range $r = \infty$. Figure 5.7a shows the exit probability as a function of the initial proportion of agents with opinion a for values of $\rho_b \in \{0.5, 0.875, 0.96875, 1.0\}$. We compare the results from the multi-agent simulations (symbols) with the numerical solutions of the master equation approximated with the Gillespie algorithm (lines). We find a good agreement between the numerical solutions of the master equation and the multi-agent simulations. For equal opinion qualities (i.e., $\rho_b = 1$), the exit probability E_N resembles a straight line with slope 1. When the difference in

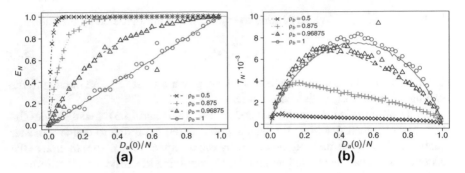

Fig. 5.7 Illustration of the effects of the quality of each option of the decision-making problem on the performance of the swarm. **a** shows the exit probability E_N as a function of the initial conditions; **b** shows the consensus time T_N. Symbols represent the results of multi-agent simulations and lines represent the results of Algorithm 5.1. Parameters: $N = 100$, $g = 100$ s, $\sigma^{-1} = 10$ s, $\rho_a = 1$, and $\rho_b \in \{0.5, 0.875, 0.96875, 1.0\}$

quality between the two options increases (i.e., $\rho_a > \rho_b$) E_N increases as well and eventually converges to a step function. This result holds for all initial conditions $\{D_a(0) \in (0, N], D_b(0) = N - D_a(0), E_a(0) = 0, E_b(0) = 0\}$. The DMVD strategy enables a swarm of agents to easily discriminate an inadequate site from a good one, while it correctly generates an unbiased behavior for sites of equal qualities.

Additionally, as shown in Fig. 5.7b, the chemical reaction network model provides a good approximation of the consensus time T_N with only a small prediction error (i.e., the difference between lines and symbols n Fig. 5.7b) that increases with increasing values of T_N. This prediction error is a result of the fact that our model neglects the delays due to traveling times. When T_N is maximal, agents perform many visits to the sites and the effects of traveling times is more pronounced. Nonetheless, traveling times only affect the transient dynamics of the systems, and therefore the consensus time (as shown in Fig. 5.7b). In contrast, equilibrium dynamics given by the exit probability are independent of such delays (see Fig. 5.7a). The DMVD strategy requires longer times to discriminate between sites of similar qualities, while easier decision-making problems are solved with much smaller effort. Note that this behavior is opposed to that of the IMMD strategy described in Chap. 4 which takes longer times to solve easier decision-making problems. Furthermore, for equal sites' quality, $\rho_a = \rho_b$, the consensus time converge to a curve that is symmetric and centered around the unbiased initial condition $\{D_a(0) = N/2, D_b(0) = N/2, E_a(0) = 0, E_b(0) = 0\}$ (which is also the case for the IMMD strategy).

5.3.2   Scalability with the Size of the Swarm

When the swarm size N is finite, the dynamics of the DMVD strategy is not deterministic and shows stochastic behavior that varies as a function of N. In the following

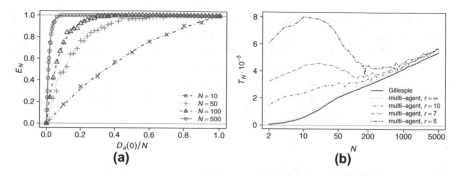

Fig. 5.8 Illustration of the effects of the size of the swarm on the accuracy and on the time necessary to the swarm for a collective decision. **a** shows the exit probability E_N as a function of the initial conditions for $N \in \{10, 50, 100, 500\}$; symbols represent the results of multi-agent simulations and *lines* represent the results of Algorithm 5.1. **b** shows the consensus time T_N as function of the swarm size N for values of the interaction range $r \in \{5, 7, 10, \infty\}$; the *solid line* represents the results of the Gillespie algorithm and the *dashed lines* represent the results of multi-agent simulations. Parameters: $g = 100\,\text{s}$, $\sigma^{-1} = 10\,\text{s}$, $\rho_a = 1$, and $\rho_b = 0.875$

we study the scalability of the DMVD strategy using the chemical reaction network model and we compare its prediction with the results of multi-agent simulations.

We consider a swarm with unlimited interaction range $r = \infty$ that is tasked with solving a site-selection problem defined by the option qualities $\rho_a = 1$ and $\rho_b = 0.875$. We use the accuracy of the decision given by the exit probability and the consensus time as performance measures. Figure 5.8a shows the exit probability E_N as a function of the initial proportion of agents favoring opinion a for different values of the swarm size N. The multi-agent simulations (symbols) and the chemical reaction network (lines) show good agreement. When the value of N is small (e.g., $N = 10$) the exit probability approaches a straight line with slope 1, resembling the initial proportion of agents with opinion a. However, as the swarm size increases, the exit probability rapidly grows and approaches a step function. Therefore, the accuracy of the DMVD strategy depends positively on the size of the swarm: bigger swarms are more accurate. Additionally, we note that the results of the chemical reaction network are in agreement with the deterministic behavior of the swarm (i.e., consensus on opinion a) predicted by the ODE model in Sect. 5.2.

Additionally, we investigate the scalability of the DMVD strategy in terms of the time T_N necessary to the swarm for a collective decision. For unlimited interaction range $r = \infty$, the multi-agent simulations are correctly predicted by the chemical reaction network as can be observed by comparing the trends of the dashed lines with that of the solid line shown in Fig. 5.8b. The small prediction error of the chemical reaction network model is a result of neglecting the traveling time necessary to the agents to move between the nest and the sites. Note that according to our model the consensus time scales logarithmically with the swarm size N. In contrast, when the interaction range is finite (e.g., $r = 5$), the chemical reaction network fails to accurately predict the consensus time for swarms of small size N. This discrepancy is reduced for larger swarm size N and larger interaction range (e.g., $r \in \{7, 10\}$). While a finite interaction range r affects the consensus time, the prediction error is reduced for larger swarms as higher agent densities reduce the effects of small r. For these finite interaction ranges r, bigger swarms have faster decision-making processes as also observed in honeybee swarms (Schaerf et al. 2013). Contrary to the accuracy of the collective decision, the consensus time is affected by both the size of the swarm and the value of the interaction range.

5.3.3   Robustness to Noisy Qualities

In real-world applications, the agents of a self-organizing swarm are most likely equipped with low-cost sensors that suffer from noisy measurements. In these conditions, the accuracy of the collective decision and the duration of the decision-making process might be affected, resulting in the loss of performance and in increased prediction errors. In this section, we study the robustness of the DMVD strategy to noisy estimations of the site qualities by agents in the exploration states. For this purpose, we include in the multi-agent simulations a normally distributed noise with zero mean and varying standard deviation $\epsilon \in \{0.05, 0.10, 0.15, 0.2\}$ over the values of the site qualities $\rho_a = 1.0$ and $\rho_b = 0.96875$.

Figure 5.9a shows the exit probability E_N for a swarm of agent with limited interaction range $r = 5$ for various levels of noise $\epsilon \in \{0.05, 0.1, 0.15, 0.2\}$. The exit probability is equivalent for all tested levels of noise and consistently predicted by the Gillespie algorithm. As observed for honeybee swarms (Passino and Seeley 2006; Schaerf et al. 2013), the DMVD strategy shows a very high robustness to noisy estimates of the quality of the sites. Figure 5.9b shows the time T_N necessary to reach consensus under noisy conditions for swarms of different size and agents

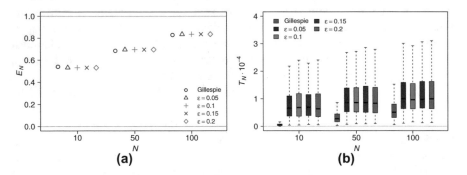

Fig. 5.9 Illustration of the effects of noisy quality estimations on the accuracy and on the time necessary to the swarm for a collective decision. **a** shows the exit probability E_N and **b** shows the consensus time T_N for varying level of noise $\epsilon \in \{0.05, 0.1, 0.15, 0.2\}$. Parameters: $g = 100\,\mathrm{s}$, $\sigma^{-1} = 10\,\mathrm{s}$, $\rho_a = 1$, $\rho_b = 0.96875$, and $r = 5$

with finite interaction range $r = 5$. The consensus time seems not to be significantly influenced by the noise. In agreement with the results in Sect. 5.3.2, the numerical approximation of the master equation fails to predict the consensus time for a swarm with limited interaction range r. Nonetheless, the prediction error decreases with increasing swarm size N and consequently our predictions are expected to be accurate for large swarms.

5.4 Discussion

In this chapter, we focused our attention on a collective decision-making strategy that combines a simple individual decision-making mechanism—the voter model—with a mechanism for the direct modulation of positive feedback. We called this strategy *Direct Modulation of Voter-based Decisions* (DMVD). The primary advantages of this collective decision-making strategy are (i) its increasing decision accuracy with increasing size of the swarm (see Fig. 5.8a), (ii) the fact that the time necessary to reach consensus scales logarithmically with the swarm size (see Fig. 5.8b), and (iii) the high robustness of the strategy to noisy estimates of the sites' quality by individual agents (see Fig. 5.9). We have reported a deterministic ODE model and a stochastic master equation model defined as a chemical reaction network. These models allow us to accurately predict the performance of the swarm in terms of the accuracy of the collective decision and the time necessary for achieving consensus. Using the ODE model we are able to guarantee convergence to a collective decision for the best option in the thermodynamic limit. Using the Gillespie simulations of the chemical reaction network we are able to give guarantees for the decision accuracy and the consensus time while accounting for finite-size effects. We empirically investigated the robustness of the DMVD strategy using multi-agent simulations and validated the accuracy of our mathematical models. We conjecture that the robustness of the DMVD strategy is a result of the distributed nature of the collective decision-making process.

The DMVD strategy is related to the approach proposed by Parker and Zhang (2009). In their study, the authors proposed a strategy to tackle the best-of-n problem that closely resembles the actual house-hunting behavior of ant colonies (Mallon et al. 2001). Agents directly recruit their neighbors as observed in ant colonies. In contrast, the DMVD strategy is based on an individual decision-making mechanism that is inspired by the waggle dance behavior of honeybees (von Frisch 1967) and performs indirect recruitment through neighbor observations. These two individual decision-making mechanisms are known to produce equivalent dynamics (Franks et al. 2002) although they might differ in cognitive requirements. The strategy proposed by Parker and Zhang (2009) includes, in addition to the deliberation phase equivalent to our dissemination states, an initial search phase where robots search for possible alternatives and a final commitment phase where the swarm recognizes that a collective decision has been made. As a consequence, the resulting control algorithm is more complex (due to the many control states) and has been studied only empirically by means of experiments with robot swarms of up to 15 robots (Parker and Zhang 2009). Later, the authors extended their study by focusing on the primary component of their algorithm that provides the swarm with discrimination capabilities and proposed a rate equation model to deepen their study of the resulting dynamics (Parker and Zhang 2011).

With respect to the indirect modulation of majority-based decisions (IMMD) strategy investigated in Chap. 4, the DMVD strategy makes use of the voter model (Clifford and Sudbury 1973; Liggett 1999) as the individual decision-making mechanism rather than the majority rule (Galam 2008). The two decision-making mechanisms provide different compromises in terms of the speed and the accuracy of the decision-making process that we will investigate in the next chapter. Additionally, the DMVD strategy also differs from the IMMD strategy in the mechanism used for the modulation of positive feedback. In the DMVD strategy, the direct modulation of positive feedback is the result of agents disseminating their opinions for a time proportional to the opinions' quality (i.e., an internal preference factor). In the IMMD strategy (cf. Chap. 4), the modulation of positive feedback is an indirect effect of an environmental bias factor, i.e., the different traveling times between regions of the environment that are associated with each opinion. The longer the traveling time the lower is the opinion quality. As a consequence, the IMMD strategy takes more time to discriminate between option of very different quality (cf. the discussion in Sect. 4.3.2) contrary to the DMVD strategy which solves such easier decision-making problems faster (see Fig. 5.7b).

References

P. Clifford, A. Sudbury, A model for spatial conflict. Biometrika **60**(3), 581–588 (1973)

G. Deffuant, D. Neau, F. Amblard, G. Weisbuch, Mixing beliefs among interacting agents. Adv. Complex Syst. **3**(01n04), 87–98 (2000)

N.R. Franks, S.C. Pratt, E.B. Mallon, N.F. Britton, D.J.T. Sumpter, Information flow, opinion polling and collective intelligence in house-hunting social insects. Philos. Trans. R. Soc. B Biol. Sci. **357**(1427), 1567–1583 (2002)

S. Galam, Sociophysics: a review of Galam models. Int. J. Modern Phys. C **19**(03), 409–440 (2008)

S. Garnier, J. Gautrais, G. Theraulaz, The biological principles of swarm intelligence. Swarm Intell. **1**(1), 3–31 (2007)

D.T. Gillespie, Exact stochastic simulation of coupled chemical reactions. J. Phys. Chem. **81**(25), 2340–2361 (1977)

H. Hamann, Towards swarm calculus: urn models of collective decisions and universal properties of swarm performance. Swarm Intell. **7**(2–3), 145–172 (2013)

K. Lerman, A. Martinoli, A. Galstyan, A review of probabilistic macroscopic models for swarm robotic systems, in *Swarm Robotics*, vol. 3342, LNCS, ed. by E. Şahin, W. Spears (Springer, Heidelberg, 2005), pp. 143–152

T.M. Liggett, *Stochastic Interacting Systems: Contact, Voter and Exclusion Processes*, vol. 324, Grundlehren der mathematischen Wissenschaften (Springer, Heidelberg, 1999)

E. Mallon, S. Pratt, N.R. Franks, Individual and collective decision-making during nest site selection by the ant Leptothorax albipennis. Behav. Ecol. Sociobiol. **50**(4), 352–359 (2001)

A. Martinoli, A. Ijspeert, F. Mondada, Understanding collective aggregation mechanisms: from probabilistic modelling to experiments with real robots. Robot. Auton. Syst. **29**(1), 51–63 (1999)

A. Martinoli, K. Easton, W. Agassounon, Modeling swarm robotic systems: a case study in collaborative distributed manipulation. Int. J. Robot. Res. **23**(4–5), 415–436 (2004)

M.A. Montes de Oca, E. Ferrante, A. Scheidler, C. Pinciroli, M. Birattari, M. Dorigo, Majority-rule opinion dynamics with differential latency: a mechanism for self-organized collective decision-making. Swarm Intell. **5**, 305–327 (2011)

C.A.C. Parker, H. Zhang, Cooperative decision-making in decentralized multiple-robot systems: the best-of-n problem. IEEE/ASME Trans. Mechatronics **14**(2), 240–251 (2009)

C.A.C. Parker, H. Zhang, Biologically inspired collective comparisons by robotic swarms. Int. J. Robot. Res. **30**(5), 524–535 (2011)

K.M. Passino, T.D. Seeley, Modeling and analysis of nest-site selection by honeybee swarms: the speed and accuracy trade-off. Behav. Ecol. Sociobiol. **59**(3), 427–442 (2006)

T.M. Schaerf, J.C. Makinson, M.R. Myerscough, M. Beekman, Do small swarms have an advantage when house hunting? The effect of swarm size on nest-site selection by *apis mellifera*. J. R. Soc. Interface 10(87) (2013)

A. Scheidler, Dynamics of majority rule with differential latencies. Phys. Rev. E **83**, 031116 (2011)

A. Scheidler, A. Brutschy, E. Ferrante, M. Dorigo, The k-unanimity rule for self-organized decision-making in swarms of robots. IEEE Trans. Cybern. **46**(5), 1175–1188 (2016)

T.D. Seeley, *Honeybee Democracy* (Princeton University Press, Princeton, 2010)

O. Soysal, E. Şahin, A macroscopic model for self-organized aggregation in swarm robotic systems, in *Swarm Robotics*, vol. 4433, LNCS, ed. by E. Şahin, W.M. Spears, A.F.T. Winfield (Springer, Heidelberg, 2007), pp. 27–42

R. Toral, C.J. Tessone, Finite size effects in the dynamics of opinion formation. Commun. Comput. Phys. **2**(2), 177–195 (2007)

G. Valentini, H. Hamann, Time-variant feedback processes in collective decision-making systems: influence and effect of dynamic neighborhood sizes. Swarm Intell. **9**(2–3), 153–176 (2015)

N.G. van Kampen, *Stochastic Processes in Physics and Chemistry* (Elsevier, Amsterdam, 1992)

M. Vigelius, B. Meyer, G. Pascoe, Multiscale modelling and analysis of collective decision making in swarm robotics. PLoS ONE **9**(11), e111542 (2014)

P.K. Visscher, S. Camazine, Collective decisions and cognition in bees. Nature **397**(6718), 400 (1999)

K. von Frisch, *The Dance Language and Orientation of Bees* (Harvard University Press, Cambridge, 1967)

Chapter 6
Direct Modulation of Majority-Based Decisions

We study a collective decision-making strategy that combines a mechanism for the direct modulation of positive feedback with the majority rule used as individual decision-making mechanism. We consider the binary site-selection scenario described in the previous chapter and focus on the study of the *Direct Modulation of Majority-based Decisions*. On the one hand, this collective decision-making strategy implements a direct modulation of positive feedback with agents advertising their opinion for a time that is proportional to the assessed quality (as defined by an internal preference factor). On the other hand, the individual decision-making mechanism is implemented by the majority rule with agents changing their opinion to the one shared by the majority of the individuals in their neighborhood. We analytically compare the majority rule with the voter model by using ordinary differential equations models, showing that the majority rule achieves faster decisions at the expense of lower accuracy. This result is confirmed for finite-size systems using a second model, based on a chemical reaction network simulated numerically using the Gillespie algorithm. Using both modeling techniques, we show that the speed-accuracy trade-off of this strategy is strongly dependent on one key parameter of the system: the number of neighbors' opinions considered by individual agents when applying the majority rule.

6.1 Decision-Making Strategy

We describe the design of the *Direct Modulation of Majority-based Decisions* (DMMD) which allows a swarm of agents to discriminate between two options based on their quality as determined by an internal preference factor (cf. Chap. 2). Although the approach is general enough for an arbitrary number of options (see Chap. 3), we focus on a binary scenario to simplify the description of the DMMD strategy. We consider the same site-selection scenario studied in Chap. 5 with a symmetric

© Springer International Publishing AG 2017
G. Valentini, *Achieving Consensus in Robot Swarms*, Studies in Computational Intelligence 706, DOI 10.1007/978-3-319-53609-5_6

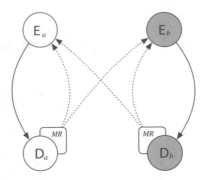

Fig. 6.1 Illustration of the probabilistic finite-state machine of the individual agent. *Solid* and *dotted lines* represent, respectively, deterministic and stochastic transitions; symbols D_i and E_i with $i \in \{a, b\}$ represent the dissemination states and the exploration states, while symbols *MR* highlight the application of the majority rule at the end of the dissemination state

environment characterized by the complete lack of environmental bias factors that might affect the modulation of positive feedback. We refer to the two options as option a and option b. The quality of the two options is denoted with $\rho_i \in (0, 1]$, $i \in \{a, b\}$. Each agent in a swarm has always a preference for an option, either a or b, referred to as the agent's opinion. Furthermore, each agent can be in one of four possible states: *dissemination* states D_a and D_b, *exploration* states E_a and E_b. The resulting probabilistic finite-state machine[1] is shown in Fig. 6.1.

The DMMD strategy uses the same mechanism for the direct modulation of positive feedback as the DMVD strategy. However, the DMMD strategy differs in the choice of the individual decision-making mechanism which is represented by the majority rule. At the end of the dissemination state (either state D_a or state D_b), the agent perceives and collects the opinions of its neighbors. Then, the agent adds its own opinion to this group of opinions and applies the majority rule to determine its next preferred option. In the remainder of this chapter, we refer to the size of this group of opinions with symbol G. Depending on the outcome of the majority rule, the agent transits to one of the two exploration states E_a or E_b (cf. dotted lines in Fig. 6.1). In the case of a tie the agent keeps its current opinion. With the exception of the individual decision-making mechanism, the implementation of the DMMD strategy corresponds to that of the DMVD strategy and the reader may refer to Chap. 5 for additional details.

---

[1]Note that the PFSM shown in in Fig. 6.1 is a specialization of the generic PFSM provided in Fig. 3.3 of Chap. 3.

6.2 Ordinary Differential Equations Model

We study the behavior of a swarm of agents executing the DMMD strategy under the continuous limit approximation ($N \rightarrow \infty$). We also study systematically the impact of the neighborhood size on the speed and accuracy of the decision-making process. To this end, we instantiate the generic system of ordinary differential equations (ODEs) defined in Eq. (3.3) of Chap. 3 and derive a specific model for the DMMD strategy. The ODE model describes the dynamics of the expected proportion of agents in the dissemination states (d_a and d_b) and the expected proportion of agents in the exploration states (e_a and e_b). Our mathematical modeling approach relies on two assumptions: (i) the neighborhood size of agents is constant and (ii) each agent has always a noiseless quality estimate of its opinion (even at time $t = 0$). Assumptions (i) and (ii) simplify our derivation of the ODE model by allowing us to neglect the effects of random fluctuations of the group size G and of the option qualities ρ_a and ρ_b, and to consider instead their mean values.

In order to define an ODE model for the DMMD strategy starting from the generic model in Eq. (3.3) we need to model the contribution of the direct modulation of positive feedback as well as that of the majority rule. As a consequence of direct modulation, the time spent in the dissemination state by an agent is proportional to the quality of the sites ($\rho_a g$ and $\rho_b g$). If these two quantities represent the average time spent by agents to disseminate their opinion, then we can also define the *rates* at which agents move from a dissemination state to an exploration state as the inverse of these quantities. We obtain, respectively, the rate $\alpha = (\rho_a g)^{-1}$ and the rate $\beta = (\rho_b g)^{-1}$. Additionally, to derive our set of differential equations, we need to know the rates at which agents change their opinions. We need to express the probability p_{ab} that an agent with opinion a switches to opinion b as an effect of applying the majority rule for a given group size G (similarly for probability p_{ba}). In the model, we also need to consider the cases where the application of the majority rule has no effect, that is, no opinion switch is triggered after its application. The probabilities of keeping the same opinion are denoted as p_{aa} and p_{bb}.

We consider a simplified example to explain how we determined these probabilities (cf. Fig. 6.2). Consider an agent i with opinion a that has two neighbors j, h.

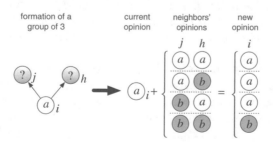

Fig. 6.2 The figure illustrates the application of the majority rule in a group of $G = 3$ agents. An agent i with opinion a applies the majority rule over a set of opinions containing its current opinion and the opinions of its two neighbors j and h. In the first three cases, agent i keeps its current preference for option a; in the last case, the agent switches its opinion to option b

Hence, we have $G = 3$. The probability p_{ab} that this agent switches opinion to b after applying the majority rule is computed by considering all possible combinations of neighbors that form a majority of opinions for option b. In this simple example with a small group, the only relevant case is when both neighbors j and h have opinion b (denoted by bb). All the other cases, aa, ab, and ba, correspond to a majority of opinions for option a which leaves agent i unaffected. We define p_a the probability that a neighboring agent has opinion a; due to symmetry, $(1 - p_a)$ is the probability that a neighboring agent has opinion b. Probability p_a is a function of the proportions d_a and d_b of agents in the dissemination states. Only these agents advertise their opinion and only they can provoke a switch, which gives $p_a = \frac{d_a}{d_a+d_b}$ (cf. Sect. 3.3). Given the probability p_a, we can derive p_{ab} as the joint probability $(1 - p_a)^2$ to have two neighbors with opinion b. In the same way, the probability p_{aa} of not provoking a switch is $p_a^2 + 2p_a(1 - p_a)$, obtained as the sum of the three cases aa, ab, and ba. The derivations of probabilities p_{ij}, i, $j \in \{a, b\}$, is performed by assuming an infinite number of agents ($N \to \infty$) and a well-mixed distribution of their positions (and therefore their opinions) within the nest. The first assumption is a direct consequence of the continuous nature of the ODE model presented in this section. The second assumption is instead motivated by the requirement of the DMMD strategy to have agents approaching a well-mixed distribution (cf. well-mixed interaction property in Sect. 3.2.1).

The above reasoning to compute probabilities p_{aa} and p_{ab} for a pair of neighbors can be generalized to a generic group size G by means of equations

$$p_{aa} = \sum_{i=\lfloor (G-1)/2 \rfloor}^{G-1} \binom{G-1}{i} p_a^i (1 - p_a)^{G-1-i}, \tag{6.1}$$

$$p_{ab} = \sum_{i=0}^{\lfloor (G-1)/2 \rfloor -1} \binom{G-1}{i} p_a^i (1 - p_a)^{G-1-i}. \tag{6.2}$$

These equations are a discrete integration of a Binomial distribution, where p_a is the success probability, $G - 1$ the number of trials, and i the number of successes. The rationale is simple. In order to keep opinion a, the number of successes for a needs to be less than half of the neighborhood $(G - 1)$ minus 1 which is the agent itself. More successes than that provoke a switch towards opinion b. The expressions for probabilities p_{bb} and p_{ba} can be obtained by swapping the power indexes in Eqs. (6.1) and (6.2).

Finally, by means of rates α and β and of probabilities p_{aa}, p_{ab} p_{ba}, and p_{bb}, we can instantiate the generic system of ordinary differential equation provided in Eq. (3.3). With the exception of probabilities p_{aa}, p_{ab} p_{ba}, and p_{bb}, the resulting ODE model is equivalent to that provided in Eq. (5.1) for the DMVD strategy. The reader may refer to the description of Fig. 5.4 provided in Chap. 5 for a detailed explanation of the flows of proportions of agents that characterize the ODE model.

6.2.1 Stability of Equilibria

Our objective in this section is to understand which are all the possible collective decisions that might emerge from the execution of the DMMD strategy. To reach this objective, we determine what are all the possible fixed points $\gamma^\star = [d_a^\star, d_b^\star, e_a^\star, e_b^\star]^T$ of the system of ODEs and perform a stability analysis. The analysis results in three fixed points

$$\gamma_1^\star = \left[\frac{g\sigma\rho_a}{1 + g\sigma\rho_a}, 0, \frac{1}{1 + g\sigma\rho_a}, 0 \right]^T, \tag{6.3}$$

$$\gamma_2^\star = \left[0, \frac{g\sigma\rho_b}{1 + g\sigma\rho_b}, 0, \frac{1}{1 + g\sigma\rho_b} \right]^T, \tag{6.4}$$

$$\gamma_3^\star = \frac{1}{\Psi} \left[g\sigma\rho_a\rho_b^2, g\sigma\rho_a^2\rho_b, \rho_b^2, \rho_a^2 \right]^T, \tag{6.5}$$

where $\Psi = \rho_a^2 + g\sigma\rho_a^2\rho_b + g\sigma\rho_a\rho_b^2 + \rho_b^2$ is a normalization constant.

The two fixed points $\gamma_1^\star$ and $\gamma_2^\star$ given by Eqs. (6.3) and (6.4) represent consensus on opinion a and consensus on opinion b. In addition to the options' qualities ρ_i, $i \in \{a, b\}$, the proportion of agents in the exploration and the dissemination states predicted by $\gamma_1^\star$ and $\gamma_2^\star$ depend also on the exploration and dissemination rates. This result means that the designer has a tool to fine-tune the desired proportion of agents exploring or disseminating at consensus. This could be of interest during a foraging task (Montes de Oca et al. 2011; Valentini et al. 2013; Scheidler et al. 2016) to effectively tune the foraging rate, or to aid the calibration of the quorum thresholds (Parker and Zhang 2009, 2010) when the detection of consensus is necessary to trigger a change in the behavior of the entire swarm (e.g., migration to the selected site). The third equilibrium $\gamma_3^\star$ in Eq. (6.5) corresponds instead to a macroscopic state of indecision where both opinions coexist in the swarm.

A subsequent question that arises is which of these equilibria is asymptotically stable and, more importantly, under which conditions. To answer this question, we linearize the system of ODEs around each equilibrium, calculate the eigenvalues of the corresponding Jacobian matrix, and study their signs. Note that, due to the conservation of the swarm mass, the system of ODEs is over-determined. One equation can be omitted, for example the last equation, by rewriting the remaining three using the substitution $e_b = 1 - d_a - d_b - e_a$. Therefore, each equilibrium of the system has only three meaningful eigenvalues. The eigenvalues corresponding to the two consensus equilibria $\gamma_1^\star$ and $\gamma_2^\star$ are

$$\begin{bmatrix} -\frac{1}{g\rho_b} \\ -\sigma \\ \frac{-g\sigma\rho_a\rho_b - \rho_a}{g\rho_a\rho_b} \end{bmatrix}, \begin{bmatrix} -\frac{1}{g\rho_a} \\ -\sigma \\ \frac{-g\sigma\rho_a\rho_b - \rho_a}{g\rho_a\rho_b} \end{bmatrix}, \tag{6.6}$$

respectively for consensus on option a and for consensus on option b. These eigenvalues depend only on the rates g, σ and on the site qualities ρ_a, ρ_b. Given that these quantities are defined to be always strictly positive we can observe that the eigenvalues in Eq. (6.6) are always strictly negative and conclude that the two consensus equilibria are asymptotically stable.

The third equilibrium γ_3^* is characterized by eigenvalues with a very complex analytic formulation which prevents us from providing it here for the reader.[2] Nonetheless, we have performed the stability analysis for this fixed point as well. According to our analysis, for values of ρ_a, $\rho_b \in (0; 1]$, $\rho_a \geq \rho_b$, and for σ, $g > 0$, two eigenvalues are always strictly negative while one is always strictly positive. Such a fixed point, which is difficult to visualize due to the high dimensionality of the system, is a saddle point and divides the basin of attraction between trajectories converging to consensus on opinion a and trajectories converging to consensus on opinion b (see also the next section). Additionally, trajectories that originate close to this fixed point are characterized by long transient dynamics (i.e., higher consensus time) before converging on either γ_1^* or γ_2^*. We can therefore conclude that the macroscopic state of indecision, γ_3^*, is not stable.

We finally compare the dynamics of the DMMD strategy with those of the DMVD strategy analyzed in Chap. 5. Let us recall that these two collective decision-making strategies differ only in the individual decision-making mechanism: the DMMD strategy implements the majority rule while the DMVD implements the voter model. With respect to the majority rule, the voter model has simpler asymptotic dynamics as it has only 2 equilibria corresponding to the two consensus decisions. One of the two equilibria is associated with the best option a and is asymptotically stable when $\rho_a > \rho_b$. The other equilibrium is unstable. When $\rho_a = \rho_b$, one eigenvalue vanishes for both equilibria that are in this case only Lyapunov stable but not asymptotically stable. Under these conditions, the voter model does not converge to a collective decision but remains indefinitely with the same proportion of agents with opinions a and b with which the swarm was initialized. Therefore, in the limit of $N \to \infty$, the differences between the voter model and the majority rule are the following. (1) With the majority rule, convergence to a particular equilibrium depends on the initial conditions (as there are two stable equilibria); whereas the voter model always converges to the best option, if one exists ($\rho_a > \rho_b$). (2) The majority rule converges, differently from the voter model, to one of the opinions even in the case of symmetric qualities ($\rho_a = \rho_b$). In Chap. 5, we showed that these properties of the voter model hold only in the deterministic, continuous approximation ($N \to \infty$), and that they vanish when the influence of finite-size effects is included (see Sect. 5.3).

---

[2]The reader might refer to the supplementary material of (Valentini et al. 2016) available online (Valentini et al. 2015) for a Mathematica notebook containing the derivation and the symbolic analysis of γ_3^*.

6.2.2 Speed Versus Accuracy Trade-Off

Our aim in this section is to use the ODE model defined above to analyze how convergence speed and decision accuracy (Franks et al. 2003; Passino and Seeley 2006) change as a function of a key parameter of our strategy: the group size G. In our terminology, the system has higher accuracy when it can reach consensus on the opinion associated to the best option (i.e., option a) for a wider range of initial conditions. For all possible initial conditions $d_a(0) \in [0, 1]$, $d_b = 1 - d_a(0)$, we determine the consensus $d_a^\star + e_a^\star \in \{0, 1\}$ that is reached asymptotically from there. We are particularly interested in the border c that separates the two basins of attraction: we converge to $d_a^\star + e_a^\star = 1$ for $d_a(0) \in [c + \varepsilon, 1]$ and to $d_a^\star + e_a^\star = 0$ for $d_a(0) \in [0, c - \varepsilon]$ where $\varepsilon > 0$. Smaller values of c are preferred since they increase the basin of attraction for the best option. We consider the convergence time as the time necessary to reach consensus on any option. To compute this time from the ODE model, we introduce a threshold $\delta = 10^{-3}$ and consider that the system has converged to a collective decision at a certain time t if either $d_a(t) + e_a(t) \geqslant 1 - \delta$ (i.e., consensus on opinion a) or $d_a(t) + e_a(t) \leqslant \delta$ (i.e., consensus on opinion b). We define convergence time to be the minimum t satisfying this criterion.

The results of this analysis are reported in Fig. 6.3 for decision accuracy and Figs. 6.4 and 6.5 for convergence time. In both figures, the difference between the left and right graphs is the value of the quality parameter ρ_b. We keep the quality of opinion a constant with $\rho_a = 1$ and vary the value of ρ_b which determines the difficulty of the decision-making problem. Specifically, a quality of $\rho_b = 0.5$ defines a simpler, more asymmetric best-of-2 problem where option a is twice as good as option b, whereas $\rho_b = 0.9$ defines a much harder problem where the qualities of the two options are more difficult to distinguish.

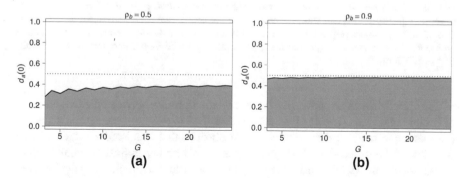

Fig. 6.3 The figure shows the results from the speed versus accuracy analysis performed using the ODE model as a function of the group size G, the initial condition $d_a(0)$, $d_b(0) = 1 - d_a(0)$, and the option quality ρ_b. The figures show the border c (*black line*) that divides initial conditions that lead the system to consensus on a (*white area*) from those with consensus on b (*gray area*) for increasing values of the group size G. **a** shows the results for $\rho_b = 0.5$ and **b** shows the results for $\rho_b = 0.9$. Parameters: $g = 8.4$ min, $\sigma^{-1} = 6.072$ min

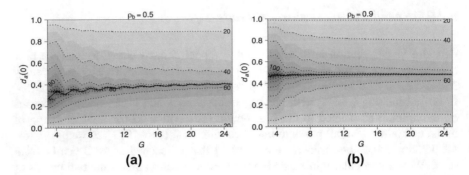

Fig. 6.4 The figure shows the results from the speed versus accuracy analysis performed using the ODE model as a function of the group size G, the initial condition $d_a(0)$, $d_b(0) = 1 - d_a(0)$, and the option quality ρ_b. The heatmaps show the consensus time (in min) for group size $G \in \{3, 25\}$ and initial condition $d_a(0) \in [0, 1]$, respectively, **a** for $\rho_b = 0.5$ and **b** for $\rho_b = 0.9$. *Black solid lines* represent the border points c for each value of G. Parameters: $g = 8.4$ min, $\sigma^{-1} = 6.072$ min

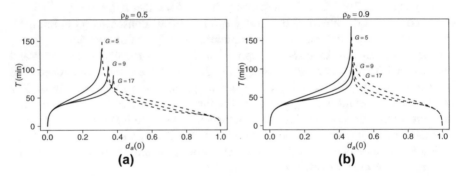

Fig. 6.5 The figure shows the results from the speed versus accuracy analysis performed using the ODE model as a function of the group size G, the initial condition $d_a(0)$, $d_b(0) = 1 - d_a(0)$, and the option quality ρ_b. Figures show the consensus time over initial conditions $d_a(0)$ for group size $G \in \{5, 9, 17\}$, respectively, **a** for $\rho_b = 0.5$ and **b** for $\rho_b = 0.9$. *Lines* represent initial conditions with consensus on option a (*dashed lines*) and option b (*solid lines*). Parameters: $g = 8.4$ min, $\sigma^{-1} = 6.072$ min

In Fig. 6.3a the black, solid line represents the border c between the two intervals of initial conditions (basins of attraction) that lead to different consensus decisions (i.e., asymptotically stable solutions). We observe that this border increases as a function of the group size G. Higher values of this border indicate a smaller set of initial conditions (white area) that lead the swarm to choose the best option (i.e., site a), and thus lower decision accuracy. The graph shows that the accuracy of the DMMD strategy decreases as a function of the group size. This happens for both easier ($\rho_b = 0.5$, Fig. 6.3a) and more difficult ($\rho_b = 0.9$, Fig. 6.3b) decision-making problems. However, for $\rho_b = 0.9$ this increase is much less noticeable due to the fact that accuracy is already relatively low for small group sizes. Additionally, we can observe that the parity of the group size G influences the accuracy of the decision-making process. When G is even the set of initial conditions leading to consensus on option a is smaller than that of the two nearby odd group sizes. This phenomenon, which is more distinct for small group sizes, is characteristic of the majority rule and was reported previously for other systems too (Galam 1986; List 2004).

Figure 6.4a, b show through heatmaps how the time necessary to reach a decision varies as a function of the group size G and of initial conditions $d_a(0), d_b(0) = 1 - d_a(0)$. The black lines provide the border c between consensus on opinion a and consensus on opinion b. As we can see, the consensus time increases with the proximity to the border c. Figure 6.5a, b detail instead the shape of consensus time for selected values of the group size G. The type of the lines represents the asymptotic result of the decision-making process, respectively, dashed lines for consensus on opinion a and solid lines for consensus on opinion b. As we can see, the consensus time is higher when the initial proportion $d_a(0)$ of agents favoring option a is closer to the border c between the basins of attraction that divides initial conditions leading to consensus on opinion a from those leading to consensus on opinion b (i.e., where lines turn from solid to dashed in Fig. 6.5a, b). Additionally, we observe that increasing the group size G speeds up the decision-making process for a wide range of initial conditions $d_a(0)$. This speedup obtained by increasing the group size becomes smaller every time we double the number of neighbors in the group (c.f. the speedup given by $G = 9$ with respect to $G = 5$ with that given by $G = 17$ with respect to $G = 9$).

The results given in Figs. 6.3, 6.4, and 6.5 reveal the crucial trade-off between convergence speed and decision accuracy of the DMMD strategy. We can increase convergence speed by increasing the group size G at the cost of lower accuracy. Similarly, we can have higher accuracy at the cost of lower convergence speed. This behavior is particularly evident for simple decision-making problems (e.g., $\rho_b = 0.5$). For more difficult best-of-2 problems (e.g., $\rho_b = 0.9$), the group size G has a lower influence on the decision accuracy while the swarm can still benefit in terms of convergence speed.

6.3 Chemical Reaction Network Model

In Sect. 6.2, we studied the asymptotic properties of the DMMD strategy using the continuous limit approximation ($N \to \infty$). As done in Sect. 5.3 for the DMVD strategy, we deepen our understanding of the DMMD strategy by performing an analysis of finite-size effects resulting from random fluctuations.

Given a swarm of N agents, we use symbols D_a and D_b to denote the number of agents in the dissemination states and symbols E_a and E_b to denote the number of agents in the exploration states. Additionally, we refer to an individual agent in one of these states using symbols $\mathfrak{D}_a$ and $\mathfrak{D}_b$ for opinion dissemination and symbols $\mathfrak{E}_a$ and $\mathfrak{E}_b$ for exploration, respectively. The DMMD strategy is modeled by the chemical reactions

---

Algorithm 6.1: Gillespie algorithm for the DMMD strategy. Parameters: $(N, G, D_a(0), \sigma, \alpha, \beta)$.

---

1 Initialize agents in the swarm: $D_a = D_a(0)$, $D_b = N - D_a(0)$, $E_a = 0$, $E_b = 0$

2 Initialize time: $t = 0$

3 **repeat**

4   Compute the total reaction rate: $\kappa = \alpha D_a + \beta D_b + \sigma E_a + \sigma E_b$

5   Generate an exponentially distributed time t' with rate parameter κ and set $t = t + t'$

6   Set $p_{aa} = \sum_{i=\lfloor(G-1)/2\rfloor}^{G-1} \binom{D_a-1}{i}\binom{D_b}{G-i-1}/\binom{D_a+D_b}{G-1}$ and
     $p_{ba} = \sum_{i=0}^{\lfloor(G-1)/2\rfloor-1} \binom{D_a}{G-i-1}\binom{D_b-1}{i}/\binom{D_a+D_b}{G-1}$

7   Let x be a uniformly distributed random number in the range $[0, 1)$

8   **if** $x \in [0; p_{aa}\alpha D_a/\kappa)$ $\longrightarrow$ Set $D_a = D_a - 1$ and $E_a = E_a + 1$

9   **if** $x \subset [p_{aa}\alpha D_a/\kappa; \alpha D_a/\kappa)$ $\longrightarrow$ Set $D_a = D_a - 1$ and $E_b = E_b + 1$

10   **if** $x \in [\alpha D_a/\kappa, \alpha D_a/\kappa + p_{ba}\beta D_b/\kappa)$ $\longrightarrow$ Set $D_b = D_b - 1$ and $E_a = E_a + 1$

11   **if** $x \in [\alpha D_a/\kappa + p_{ba}\beta D_b/\kappa; \alpha D_a/\kappa + \beta D_b/\kappa)$ $\longrightarrow$ Set $D_b = D_b - 1$ and $E_b = E_b + 1$

12   **if** $x \in [\alpha D_a/\kappa + \beta D_b/\kappa; \alpha D_a/\kappa + \beta D_b/\kappa + \sigma E_a/\kappa)$ $\longrightarrow$ Set $E_a = E_a - 1$ and $D_a = D_a + 1$

13   **if** $x \in [\alpha D_a/\kappa + \beta D_b/\kappa + \sigma E_a/\kappa; 1)$ $\longrightarrow$ Set $E_b = E_b - 1$ and $D_b = D_b + 1$

14 **until** $D_a + E_a \in \{0, N\}$;

$$\mathfrak{D}_a \xrightarrow{\alpha} \mathfrak{E}_a | \mathfrak{E}_b, \tag{6.7}$$

$$\mathfrak{D}_b \xrightarrow{\beta} \mathfrak{E}_a | \mathfrak{E}_b, \tag{6.8}$$

$$\mathfrak{E}_a \xrightarrow{\sigma} \mathfrak{D}_a, \tag{6.9}$$

$$\mathfrak{E}_b \xrightarrow{\sigma} \mathfrak{D}_b. \tag{6.10}$$

The above set of reactions is sufficient to define a master equation as described by van Kampen (1992). According to Eqs. (6.7) and (6.8), each molecule representing an agent in a dissemination state (either $\mathfrak{D}_a$ or $\mathfrak{D}_b$) transforms into a molecule representing an agent in the exploration state (either $\mathfrak{E}_a$ or $\mathfrak{E}_b$) at a constant rate. Specifically, at rate $\alpha = (\rho_a g)^{-1}$ if the agent is in state D_a or at rate $\beta = (\rho_b g)^{-1}$ otherwise. Equations (6.9) and (6.10) model instead the transformation of molecules representing agents in the exploration state (either $\mathfrak{E}_a$ or $\mathfrak{E}_b$) into molecules representing agents in the dissemination state (either $\mathfrak{D}_a$ or $\mathfrak{D}_b$) which happens at a constant rate σ.

In the Gillespie algorithm (Gillespie 1977), the evolution in time of the numbers of agents D_a, D_b, E_a, and E_b is obtained by iteratively performing two steps: (i) determine the time of the next reaction and (ii) determine which is the reaction that occurs and consequently update the macroscopic state D_a, D_b, E_a, E_b of the system. Since the execution time of chemical reactions is modeled by exponentially distributed times (van Kampen 1992), we have that the time before the next occurrence of any reaction is also exponentially distributed. Specifically, this time is computed

as the minimum of a set of exponentially distributed variables which is still exponentially distributed with a rate κ equal to the sum of the individual reactions rates
(see lines 4–5 in Algorithm 6.1). The specific reaction that occurs is randomly determined with probabilities equal to the ratio between each reaction rate and the overall
rate κ. If the reaction in Eq. (6.7) occurs, respectively that in Eq. (6.8), the outcome
is determined by an additional probabilistic experiment. The number D_a of agents
(D_b) decreases by one unit, and the type of agents increasing by one unit is E_a with
probability p_{aa} (p_{ba}) or E_b with probability p_{ab} (p_{bb}). If the reaction in Eq. (6.9)
occurs, respectively that in Eq. (6.10), the outcome is uniquely determined. We have
that the number E_a of agents exploring option a (E_b) decreases by one unit and the
number D_a of agents disseminating opinion a (D_b) increases by one unit.

This additional step is required because the reactions in Eqs. (6.7) and (6.8) are
in fact "meta-reactions" that expand in a larger reaction set having one entry for
each possible configuration of an agent neighborhood during the application of the
majority rule. Probabilities p_{aa}, p_{ab}, p_{ba}, and p_{bb} are the discrete equivalent of the
switching probabilities in Eqs. (6.1) and (6.2) used for the continuous ODE model.
In contrast to the Binomial distribution used in Sect. 6.2, in the discrete case we use
an hypergeometric distribution, which yields probabilities

$$p_{aa} = \sum_{i=\lfloor (G-1)/2 \rfloor}^{G-1} \frac{\binom{D_a-1}{i}\binom{D_b}{G-i-1}}{\binom{D_a+D_b}{G-1}}, \tag{6.11}$$

$$p_{ab} = \sum_{l=0}^{\lfloor (G-1)/2 \rfloor -1} \frac{\binom{D_a-1}{i}\binom{D_b}{G-i-1}}{\binom{D_a+D_b}{G-1}}. \tag{6.12}$$

Probabilities p_{aa} and p_{ab} are a discrete integration of an hypergeometric distribution.
Using the standard terminology of the hypergeometric distribution, we have that D_a
and D_b are the number of success states and failure states in the population, $D_a + D_b$
is the population size, $G - 1$ is the number of trials, and i the actual number of drawn
successes. The expressions for probabilities p_{bb} and p_{ba} can be obtained by swapping
the number of successes i with the number of failures $G - i - 1$ in Eqs. (6.11) and
(6.12).

We use the Gillespie algorithm to simulate a number of trajectories of the chemical
reaction network and use them to compute the exit probability E_N, i.e., the probability
that a swarm of N agents reaches consensus on opinion a, and the average consensus
time T_N, i.e., the time necessary to reach consensus on any option. In all of our studies,
we approximate the values of E_N and T_N using 2.5×10^4 independent executions of
Algorithm 6.1 for each data point. In the reminder of this section, we use the chemical
reaction network model to perform a thorough analysis of the speed versus accuracy
trade-off that characterizes the DMMD strategy and we compare the results with
those of the DMVD strategy previously described in Chap. 5.

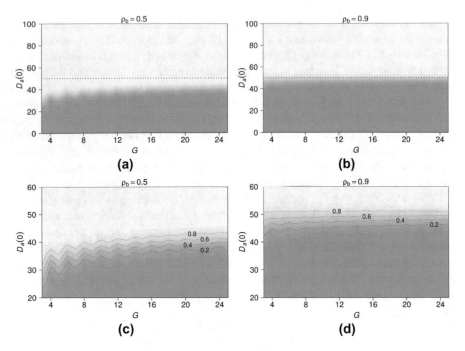

Fig. 6.6 Analysis of the decision accuracy using the chemical reaction network model. The x-axis refers to the group size $G \in \{3, \ldots, 25\}$, the y-axis to the initial condition $D_a(0) \in \{0, \ldots, N\}$, $D_b(0) = N - D_a(0)$. The heatmaps show the exit probability E_N to reach consensus on option a, respectively, **a** for $\rho_b = 0.5$ and **b** for $\rho_b = 0.9$. **c** and **d** are zoomed-in versions of (**a**) and (**b**). The *white color* represents consensus on opinion a ($E_N = 1$) while the *gray color* represents consensus on opinion b ($E_N = 0$). Parameters: $g = 8.4\,\text{min}$, $\sigma^{-1} = 6.072\,\text{min}$, $N = 100$

6.3.1 Speed Versus Accuracy Trade-Off

The results of the speed versus accuracy analysis performed by approximating the chemical master equation are reported in two separate figures: Fig. 6.6 reports the accuracy of the system by showing the exit probability E_N as a function of the group size G and of the initial condition $D_a(0)$; Fig. 6.7 reports the convergence speed of the DMMD strategy by showing the time T_N necessary to reach consensus as a function of the same two parameters. Throughout this analysis we keep the same color and lines schema used in the figures of Sect. 6.2.2 for the ODE model with the purpose of simplifying the comparison of the results.

For what concerns accuracy, the outcome of the analysis with the Gillespie algorithm is in accordance with that obtained with the ODE model: the system is more accurate for lower values of the group size G, particularly for easier decision-making problems ($\rho_b = 0.5$, Fig. 6.6a, c). The main difference between the continuous and finite-size analysis is that in the latter case we do not have anymore a clear border dividing the two basins of attraction for different consensus decisions. In contrast,

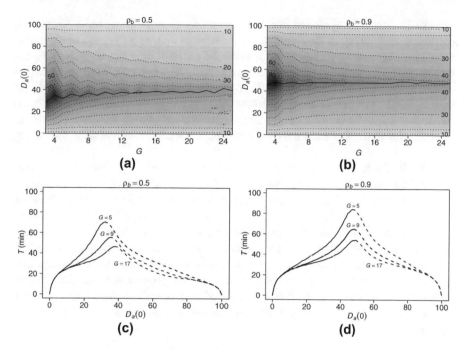

Fig. 6.7 Analysis of the convergence speed using the chemical reaction network model. The x-axis refers to the group size $G \in \{3, \ldots, 25\}$, the y-axis to the initial condition $D_a(0) \in \{0, \ldots, N\}$, $D_b(0) = N - D_a(0)$. The heatmaps show the average consensus time T_N (min), respectively, **a** for $\rho_b = 0.5$ and **b** for $\rho_b = 0.9$. *Black solid lines* represent the border points c for each value of G. In the lower row, figures show the consensus time over initial conditions $d_a(0)$ for group size $G \in \{5, 9, 17\}$, respectively, **c** for $\rho_b = 0.5$ and **d** for $\rho_b = 0.9$. *Lines* represent initial conditions with consensus on option a (*dashed lines*) and option b (*solid lines*). Parameters: $g = 8.4$ min, $\sigma^{-1} = 6.072$ min, $N = 100$

we obtain a border that gathers all the points having equal probability to converge to either of the two options ($E_N = 0.5$). Under this line, the probability to converge to option a smoothly decreases to 0, above this line, it increases to 1 (Fig. 6.6c, d). This behavior is a direct consequence of finite-size effects modeled by the chemical reaction network and ignored in the ODE model. Where the ODE model predicts a macroscopic state of indecision, we have instead that the swarm converges anyway to consensus in the finite-size model. Additionally, the results in Fig. 6.6 also show the same pattern in the decision accuracy that we observed with the ODE model where groups with even cardinality were less accurate than groups with odd cardinality (cf. Sect. 6.2.2).

The results of the analysis of the convergence speed are shown in Fig. 6.7. In agreement with the prediction obtained with the ODE model, we have that the system is faster for higher values of the group size G. This is particularly true for initial conditions that are closer to the state of indecision (i.e., the black line that marks the points with $E_N = 0.5$) as shown in Fig. 6.7a, b. The primary difference between the current finite-size analysis and that of the continuous approximation in Sect. 6.2 is evident when looking at the shape of the consensus time as a function of the initial condition. Figure 6.7c, d show these results: the curve of the consensus time T_N has a much smoother shape around the point of indecision when compared to the curves shown in Fig. 6.5a, b. Additionally, we also observe that the value of T_N predicted by the chemical reaction network is lower than the one predicted by the ODE model for almost all initial conditions. As already mentioned above, the discrepancies in the behavior of the system are a direct consequence of the finiteness of the swarm size.

Overall, the speed versus accuracy analysis presented here provides the same message as the continuous analysis in Sect. 6.2.2. By increasing the group size G, the swarm benefits in terms of convergence speed at the cost of a lower decision accuracy. This loss in accuracy is stronger in easier decision-making problems (e.g., $\rho_b = 0.5$) while it is mitigated in more difficult problems (e.g., $\rho_b = 0.9$). Conversely, the benefits concerning convergence speed obtained by increasing the group size are relatively unaffected by the difficulty of the problem. With respect to the continuous approximation provided by the ODE model, the analysis of the chemical master equations allowed us to better quantify the performance of the system by capturing the stochastic effects resulting from random fluctuation characteristic of a finite swarm.

6.3.2   Comparison with the DMVD Strategy

We conclude the analysis of finite-size effects by performing again a speed versus accuracy study but this time we compare the performance of the DMMD strategy against that of the DMVD strategy described in Chap. 5. For this purpose, we employ the chemical reaction network previously proposed in (Valentini et al. 2014) for the DMVD strategy which is derived under the same assumptions of Algorithm 6.1 and allows us to perform a fair comparison between the models. Let us recall that the only difference between the two collective decision-making strategies is given by the decision-making mechanism utilized by individual agents—in the DMVD strategy agents use the voter model and decide by copying the opinion of a random neighbor; instead, in the DMMD strategy, agents adopt the opinion favored by the majority of their neighbors according to the majority rule.

The difference $E_N^{VM} - E_N^{MR}$ in decision accuracy between the two strategies is reported in Fig. 6.8a for increasingly difficult decision-making problems ($\rho_b \in \{0.5, 0.7, 0.9, 0.99\}$). The DMMD strategy is less accurate than the DMVD strategy when values are greater than 0; both strategies are equally accurate for values equal to 0; and the DMMD strategy is more accurate than the DMVD strategy for values smaller than 0. As it can be noticed, the voter model used in the DMVD strategy is in general more accurate than the majority rule used in the DMMD strategy for initial conditions $D_a(0) \leqslant 50$. Conversely, the accuracy of the majority rule reaches that of the voter model for initial conditions $D_a(0) > 50$ and it even outperforms the voter

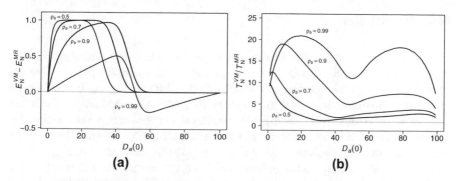

Fig. 6.8 The figure shows the results of the speed versus accuracy comparison between the majority rule and the voter model. Data shown have been generated using Algorithm 6.1 for the DMMD strategy and Algorithm 5.1 for the DMVD strategy. **a** shows the difference $E_N^{VM} - E_N^{MR}$ between the exit probability of the DMVD strategy, E_N^{VM}, and that of the DMMD strategy, E_N^{MR}. With respect to the DMVD strategy, the DMMD strategy is less accurate for $E_N^{VM} - E_N^{MR} > 0$, equally accurate for $E_N^{VM} - E_N^{MR} = 0$, and more accurate for $E_N^{VM} - E_N^{MR} < 0$. **b** shows the time ratio T_N^{VM}/T_N^{MR} between the consensus time of the DMVD strategy, T_N^{VM}, and the DMMD strategy, T_N^{MR}. Parameters: $g = 8.4\,\text{min}$, $\sigma^{-1} = 6.072\,\text{min}$, $N = 100$, $G = 5$, $\rho_b \in \{0.5, 0.7, 0.9, 0.99\}$

model for the hardest decision problem ($\rho_b = 0.99$). This behavior can be understood also in terms of the stability analysis provided in Sect. 6.2.1: since the DMVD strategy has only one asymptotically stable state, when considering finite-size effects, we have that its dynamics converge with high probability to the best option for a larger set of initial conditions (cf. Valentini et al. (2014)). In contrast, the dynamics of the DMMD strategy strongly depend on the initial conditions. Figure 6.8b reports instead the comparison between the two strategies in terms of convergence time depicted as the ratio between the consensus time of the voter model, T_N^{VM}, over that of the majority rule, T_N^{MR}. We can observe that the majority rule considerably speeds up the decision-making process for all considered parameters. The difference in speed ranges from almost two-fold for the simple problem with $\rho_b = 0.5$ up to twenty-fold for the difficult problem with $\rho_b = 0.99$.

The above analysis shows that, for application scenarios where the time available to reach a collective decision is critical, the majority rule underlying the DMMD strategy is a better design choice than the voter model used by the DMVD strategy. This key difference is extremely relevant when considering the limitations in energy autonomy of nowadays robotic platforms. Although the voter model is more accurate than the majority rule for most initial conditions $D_a(0) < 50$, this difference is considerably reduced for values of $D_a(0) \approx 50$ and vanishes for $D_a(0) > 50$. That is, when the designer has means to initialize the swarm with an approximately uniform distribution of opinion, the difference in decision accuracy between the DMMD and the DMVD strategies is negligible.

6.4 Discussion

In this chapter, we have described a collective decision-making strategy—*Direct Modulation of Majority-based Decisions* (DMMD)—that allows a swarm of agents to address the best-of-n problem. Following the DMMD strategy, individual agents in the swarm couple the use of the majority rule with a mechanism for the direct modulation of positive feedback to achieve consensus on the best option. Specifically, agents iteratively alternate periods of opinion dissemination, where they advertise their preference for particular options of the best-of-n problem, with periods of exploration, where they gather information from the environment concerning the quality of their current opinion. Similarly to (Parker and Zhang 2009, 2011; Valentini et al. 2014), the information gathered from the environment is utilized by individual agents to directly modulate their efforts of opinion promotion, that is, amplifying or reducing the time spent in the dissemination state during which they advertise a particular option. As in (Montes de Oca et al. 2011), at the end of the dissemination period, agents reconsider their current opinion by adopting the opinion favored by the majority of their neighbors. The coupling of direct modulation of the positive feedback with the majority rule implemented by the DMMD strategy introduces a bias in the dynamics of agents opinions that steers the swarm towards a collective decision for the best option.

We have defined a mathematical framework to analyze the performance of the DMMD strategy. We have investigated the limiting dynamics ($N \rightarrow \infty$) of the DMMD strategy using an ordinary differential equation model and finite-size effects ($N < \infty$) using a chemical reaction network approximated with the Gillespie algorithm. Using this mathematical framework, we proved that consensus decisions are the only asymptotically stable solutions of the system. We investigated the trade-off between the convergence speed and the decision accuracy that arises when varying the average neighborhood size of agents applying the majority rule. The primary result of this analysis is that quicker collective decisions can be obtained with larger neighborhood sizes at the cost of a lower probability to reach the optimal decision. Additionally, we observed that the parity of the group of opinions utilized in the majority rule influences this trade-off as well, with swarms using odd group sizes having greater chances to choose the best option than swarms using even groups. Finally, we compared the performance of the DMMD strategy with that of the DMVD strategy described in the previous chapter. With respect to the voter model used in the DMVD strategy, the majority rule implemented by the DMMD strategy speeds up the decision-making process considerably while it is characterized by a lower accuracy in all but the harder decision-making problems.

References

N.R. Franks, A. Dornhaus, J.P. Fitzsimmons, M. Stevens, Speed versus accuracy in collective decision making. Proc. R. Soc. B Biol. Sci. **270**, 2457–2463 (2003)

S. Galam, Majority rule, hierarchical structures, and democratic totalitarianism: a statistical approach. J. Math. Psychol. **30**(4), 426–434 (1986)

D.T. Gillespie, Exact stochastic simulation of coupled chemical reactions. J. Phys. Chem. **81**(25), 2340–2361 (1977)

C. List, Democracy in animal groups: a political science perspective. Trends Ecol. Evol. **19**(4), 168–169 (2004)

M.A. Montes de Oca, E. Ferrante, A. Scheidler, C. Pinciroli, M. Birattari, M. Dorigo, Majority-rule opinion dynamics with differential latency: a mechanism for self-organized collective decision-making. Swarm Intell. **5**, 305–327 (2011)

C.A.C. Parker, H. Zhang, Cooperative decision-making in decentralized multiple-robot systems: the best-of-*n* problem. IEEE/ASME Trans. Mech. **14**(2), 240–251 (2009)

C.A.C. Parker, H. Zhang, Collective unary decision-making by decentralized multiple-robot systems applied to the task-sequencing problem. Swarm Intell. **4**, 199–220 (2010)

C.A.C. Parker, H. Zhang, Biologically inspired collective comparisons by robotic swarms. Int. J. Robot. Res. **30**(5), 524–535 (2011)

K.M. Passino, T.D. Seeley, Modeling and analysis of nest-site selection by honeybee swarms: the speed and accuracy trade-off. Behav. Ecol. Sociobiol. **59**(3), 427–442 (2006)

A. Scheidler, A. Brutschy, E. Ferrante, M. Dorigo, The k-unanimity rule for self-organized decision-making in swarms of robots. IEEE Trans. Cybern. **46**(5), 1175–1188 (2016)

G. Valentini, M. Birattari, M. Dorigo, Majority rule with differential latency: an absorbing Markov chain to model consensus, in *Proceedings of the European Conference on Complex Systems 2012*, eds. by T. Gilbert, M. Kirkilionis, G. Nicolis. Springer Proceedings in Complexity (Springer, 2013), pp. 651–658

G. Valentini, H. Hamann, M. Dorigo. Self-organized collective decision making: the weighted voter model. in *Proceedings of the 13th International Conference on Autonomous Agents and Multiagent Systems, AAMAS'14* eds. by A. Lomuscio, P. Scerri, A. Bazzan, M. Huhns (IFAAMAS, 2014), pp. 45–52

G. Valentini, E. Ferrante, H. Hamann, M. Dorigo, Collective decision with 100 Kilobots: speed versus accuracy in binary discrimination problems (2015), http://iridia.ulb.ac.be/supp/IridiaSupp2015-005/. Accessed 24 Apr 2016

G. Valentini, E. Ferrante, H. Hamann, M. Dorigo, Collective decision with 100 Kilobots: speed versus accuracy in binary discrimination problems. Auton. Agents Multi-Agent Syst. **30**(3), 553–580 (2016)

N.G. van Kampen, *Stochastic Processes in Physics and Chemistry* (Elsevier, Amsterdam, 1992)

Part III
Robot Experiments

Chapter 7
A Robot Experiment in Site Selection

The combination of macroscopic mathematical models with simple microscopic multi-agent simulations is advantageous because it provides us with computationally efficient means to systematically analyze the dynamics of a collective decision-making strategy. However, these models rely on simplifying assumptions and might be limited in their accuracy. A validation performed through robot experiments, notwithstanding the controlled experimental conditions, provides us with extremely useful insights concerning the performance of a robot swarm. In this chapter, we aim at exploring the efficacy and the robustness of the collective decision-making strategies in real-world conditions. We consider the Direct Modulation of Majority-based Decisions applied to a binary site-selection scenario and perform a series of robot experiments using a swarm of 100 Kilobots. In doing so, we aim both at testing whether this strategy is sufficiently simple to be implemented using robots with limited hardware and at exploring its robustness to robot failures. Additionally, we compare the resulting swarm dynamics with the predictions of macroscopic mathematical models previously developed with the objective to further deepen our analysis of the speed versus accuracy trade-off characterizing these strategies.

7.1 Robotic Platform and Experimental Setup

We performed experiments using the Kilobot robotic platform (Rubenstein et al. 2014a). The Kilobot, that is shown in Fig. 7.1a, is a commercially available robot that enables researchers to experiment with large robot swarms. The Kilobot is small, with a diameter of only 3.3 cm. It has a lithium-ion battery that provides the robot with a few hours autonomy. The Kilobot moves on a smooth, two-dimensional surface using stick-slip motion. A pair of vibrating engines allow the Kilobot to move by performing micro-jumps over its three (stick-like) legs. By varying the vibration frequency of the two engines, the robot either proceeds in a straight line at a nominal speed of 1 cm/s, or turns in place at up to $\pi/4$ rad/s. The Kilobot can perceive its environment using a light sensor that allows the robot to locally measure the level

© Springer International Publishing AG 2017
G. Valentini, *Achieving Consensus in Robot Swarms*, Studies in Computational
Intelligence 706, DOI 10.1007/978-3-319-53609-5_7

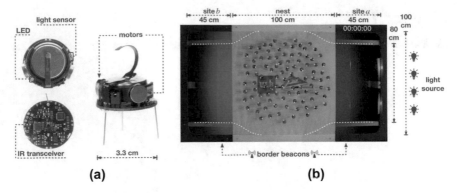

Fig. 7.1 Illustration of the robotic platform and the experimental setup. **a** shows, in clockwise order, the *top*, front and *bottom* views of the Kilobot robot highlighting the position of the ambient light sensor, the RGB LED, the motors, and the IR transceiver. **b** shows a *top*-view picture of the arena used during the robot experiments highlighting the partitioning of the environment into nest, site a and site b, as well as the positions of the border beacons and the external light source

of brightness of the ambient light. Additionally, using an infrared transceiver, the Kilobot can communicate 3-byte messages with neighbor robots up to a distance of 20 cm. Finally, the robot is endowed with one RGB LED that can be used to display internal information to an external observer.

We implemented the DMMD strategy in a binary site-selection scenario, i.e., a best-of-n problem with $n = 2$ options. We built a rectangular arena whose total size is 100×190 cm^2 (see Fig. 7.1b) which is three orders of magnitude larger than the footprint of a single Kilobot. Options a and b correspond to foraging *sites* of quality ρ_a and ρ_b, respectively. The two sites are 80×45 cm^2 large and are located at the right (site a, red) and at the left (site b, blue) side of the arena. The remaining, central part of the arena is called *nest*. It is 100×100 cm^2 large and it is where the swarm of 100 Kilobots is initially placed. The nest is also the decision-making hub of the swarm, that is, the individual decision-making mechanism is only allowed to be executed within the nest. We initially place robots in a circular area whose radius is 40 cm and whose center is the center of the nest (see Fig. 7.1b). The robots are placed so that they are approximately at the same distance from their neighbors. Their opinions are initially homogeneously distributed in the nest. At time $t = 0$, the swarm consists of 50 robots with opinion a and 50 with opinion b, all of which are initialized in the dissemination state; their initial quality estimate is unbiased ($\hat{\rho}_a(0) = \hat{\rho}_b(0) = 1$). Our goal is to have the majority of the swarm foraging from the site associated with the higher quality, in this scenario by definition site a. Specifically, the quality of site a is twice as high as that of site b ($\rho_a = 1$ and $\rho_b = 0.5$). We position a light source on the right side of the arena, to provide a landmark that can be used by the robots to navigate and find the three areas. They perform phototaxis when they need to move from site b to the nest or from the nest to site a and anti-phototaxis in the remaining two cases.

The Kilobots can identify the two sites and measure the sites' quality by using their infrared sensors. For each site, five additional Kilobots are positioned upside-down under the transparent surface of the arena, at the border between the site and the nest, and act as beacons (see Fig. 7.1b). These Kilobots continuously communicate locally a message containing the type (a or b) and the quality (ρ_a or ρ_b) associated to a site. These infrared messages are perceived only within the sites, both due to their local nature (they are positioned approximately 15 cm below the surface of the arena) and because we cover the nest area by light occluding paper to prevent robots from sensing this information at the nest.

As defined by the DMMD strategy in Chap. 6, robots continuously alternate between a period of exploration and a period of dissemination. Robots explore the site associated with their current opinion by navigating from the nest to that site and measuring its quality. They then return to the nest, where they disseminate their current opinion modulating the positive feedback based on the measured quality $\rho_i, i \in \{a, b\}$. Finally, they collect the opinions of their neighbors and apply the majority rule potentially changing preference for the best site. As explained in Sect. 6.1, the swarm can potentially suffer from the formation of clusters of robots with the same opinion, and consequently, from opinion fragmentation (Deffuant et al. 2000). For example, the robots might distribute themselves in such a way that all robots with opinion a are positioned close to site a and all robots with opinion b are positioned close to site b. As a consequence, a robot would be more likely to interact with a robot of the same opinion which might cause the decision-making process to enter a deadlock. To maintain the spatial distribution close to a well-mixed distribution, we implemented specialized motion routines that, if performed for a sufficiently long period of time, allow robots to mix well in the nest while disseminating their opinions.

7.2 Robot Control Algorithm

We implemented the DMMD strategy by using the motors, the light sensor, and the infrared transceiver of the Kilobot. Three low-level motion routines, respectively, *random walk*, *phototaxis* and *anti-phototaxis*, allow robots to navigate and to explore the environment. Depending on the current control state and on the current robot opinion, these routines are combined into a probabilistic finite-state machine to implement the behavior in the dissemination states (see Fig. 7.2a) and in the exploration states (see Fig. 7.2b). The interested reader can refer to Valentini et al. (2015b) for a video highlighting intermediate phases of the robot controller. In the following, we employ the exponential distribution to determine the duration of several sub-routines used in the robot controller. We have chosen this distribution due to its large variance that allows us to break the synchrony in the robot motion patterns by introducing noise that improves the mixing of robots.

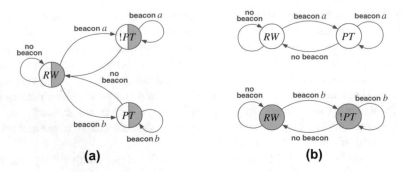

(a)   **(b)**

Fig. 7.2 The figure shows the finite-state machines that implement the motion control method of the individual robot during the execution of the decision-making strategy. **a** shows the FSM used for both dissemination states D_a and D_b; **b** shows the two FSMs used for the exploration state E_a (*top*) and E_b (*bottom*). Symbols represent low level motion routines, respectively, random walk (RW), phototaxis (PT), and antiphototaxis ($!PT$); colors represent the current robot opinion, respectively, *white* for opinion a and *gray* for opinion b

7.2.1   Low-Level Motion Routines

We implemented a correlated random walk in order to improve the mixing of the opinions in the swarm (see Table 7.1 for the parameters of this routine). When performing the random walk, the robot moves forward for a normally distributed amount of time. Given that the nest is bigger than each of the two sites, the duration of this period of time depends on the robot control state: the dissemination state is characterized by a longer period of forward motion than the exploration state. Then, the robot turns in place for a uniformly distributed period of time. Additionally, we implemented phototaxis and anti-phototaxis motion routines to allow robots in the swarm to navigate between different regions of the environment. Phototaxis is implemented by letting robots perform oriented motion towards the light source placed on the right side of the arena. The robots search for the direction with the highest light intensity by turning on the spot; once found, they move forward until the ambient light intensity measurement falls outside a tolerance range; when this happens, the robots resumes

Table 7.1 Parameters of the random walk routine. Symbol $\mathcal{N}$ represents a normal distribution with mean μ and variance σ^2, symbol $\mathcal{U}$ represent a uniform distribution. All parameters are given in seconds

Control state	Motion	Parameters
Dissemination	Forward	$\mathcal{N}(\mu = 15, \sigma^2 = 25)\,\mathrm{s}$
	Rotation	$\mathcal{U}(-3, 3)\,\mathrm{s}$
Exploration	Forward	$\mathcal{N}(\mu = 5, \sigma^2 = 9)\,\mathrm{s}$
	Rotation	$\mathcal{U}(-3, 3)\,\mathrm{s}$

the on-spot search of the correct direction of motion. Anti-phototaxis is implemented similarly to phototaxis with the only difference that, in order to move away from the light source, robots search for the direction with the lowest light intensity.

7.2.2 Dissemination States

In both dissemination states D_a and D_b, the robots execute the finite-state machine depicted in Fig. 7.2a. Robots start by performing a random walk in the nest while communicating locally their opinions. The random walk favors the spatial mixing of robots in space and therefore of their opinions. In addition to their current opinion, robots communicate a randomly generated 16-bit identifier that, with high probability, uniquely identifies the robot in its local neighborhood. This is used to make sure that, at any given time, robots distinguish the opinion of different neighbors. In general, any implementation that prevents robots from counting the opinion of a same neighbor multiple times will suffice for this purpose (see Mathews et al. 2015 for an ID-free communication example based on a combination of cameras, LED lights, and blob detection algorithms). Robots directly modulate positive feedback by spending an exponentially distributed amount of time in the dissemination state. The mean of this exponential distribution is either $\hat{\rho}_a g$ or $\hat{\rho}_b g$, where $\hat{\rho}_i$, $i \in \{a, b\}$ is the current robot estimate of the option quality. While performing the dissemination state, the robots might perceive messages from the five robot-beacons positioned at each border between the nest and a site. If such a message is perceived, it means that the robot is mistakenly leaving the nest and it therefore performs either phototaxis or anti-phototaxis in order to return to the nest (see Fig. 7.2a). Oriented motion is performed by the robot for as long as beacon messages are received and proceeds for an additional period of 20 s after the last message. This kind of oriented motion allows the robot to keep a distance from the border and to favor a good mixture of robot opinions in space. Once opinion dissemination is completed, the robot records the opinions of its neighbors for three seconds. It then adds its own current opinion to that record, applies the majority rule to determine its next preferred option and, consequently, the next site to explore. We chose a relatively short time for opinion collection in order to reduce the time-correlation of the observed opinions (i.e., robots taking decisions on the basis of outdated information). Nonetheless, this period of time is sufficient for a robot to receive messages from many neighbors as will be clear from the analysis in the next section. Finally, the robot leaves the nest to explore the chosen site.

7.2.3 Exploration States

In states E_a and E_b, robots move to the site associated with their current opinion, performing either phototaxis (towards site a) or anti-phototaxis (towards site b). Once they reach the site, they explore it for an exponentially distributed amount of time, they record the associated quality (received from the beacons), and then return to the nest. During this time, the robot executes the finite-state machine depicted in Fig. 7.2b (respectively, top for site a and bottom for site b) in order to stay within the boundaries of the site. We consider this behavior as an abstraction of a quality-estimation routine dependent on the target scenario. This abstraction allows us to study swarm dynamics that are closer to those of a real-world scenario, where exploration is a necessary and time-consuming task. For example, the robot might assess during this period how much of a certain resource is available in the site (e.g., construction material), what is the average level in the site of a certain physical feature (e.g., temperature), etc. Additionally, to ensure that robots fully enter the site (i.e., they do not remain in the border region), we implemented the following mechanism. If a robot wants to explore site a (respectively, b), it performs phototaxis (anti-phototaxis) in two phases. In the first phase, the robot performs phototaxis (anti-phototaxis) until it perceives a message from the beacons, indicating that the robot has crossed the border and entered site a (b). In the second phase, phototaxis (anti-phototaxis) is continued for as long as messages from the beacons are not received for 5 s. Exactly the same mechanism, but reversed, is used by the robots returning to the nest and entering the dissemination state. The second phase also eases the mixing of robot opinions in the nest because robots are programmed to approach the center of the nest.

7.3   Experiments

Our main working hypothesis in the analysis performed in Chap. 6 was that the efficiency and the accuracy of the DMMD strategy are affected by the neighborhood size of robots when applying the majority rule. The neighborhood size can be directly or indirectly controlled by the experimenter. However, its value could fluctuate over time due to spatial density constraints. In our scenario, we consider two extreme situations. We restrict the maximum neighborhood size to either 4 robots or to 24 robots. The latter case corresponds in practice to no restriction, since the actual number of neighbors perceived by a robot at a given time is rarely greater than 24. Robots record the opinions they receive in a memory of fixed size according to a first-in, first-out policy. Since robots receive messages from their neighbors in a random order, this implementation results in a random selection of the neighbors' opinions. We refer to this parameter as the maximum size of the opinion group G_{max} and we define it in a way so that it also includes the opinion of the considered robot: $G_{max} \in \{5, 25\}$. For each of these two cases, we performed 10 independent runs where each run lasts for 90 min.[1] Recall that the parameter g determines the duration of the dissemination state without considering the modulation of positive feedback through the site quality. The higher the value of g, the longer the robot performs its

---

[1] All experiments with the robots have been recorded and videos can be found online in the supplementary material (Valentini et al. 2015a) of the article (Valentini et al. 2016).

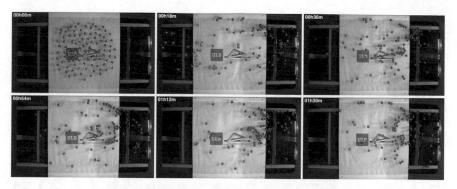

Fig. 7.3 The figure shows a series of screen-shots taken from one experiment with a swarm of 100 Kilobots. The screen-shots are taken every 18 min of execution

random walk behavior contributing to the mixing of the opinions. However, higher values of g also increase the time necessary to the swarm to find consensus. We performed preliminary test runs with parameter $g \in \{300, 400, 500\}$ s and visually evaluated the mixing of robots' opinions. We found that $g = 500$ s (i.e., about $g = 8.4$ min) provided a proper mixing of the robots' opinions while limiting the overall decision and experimentation time. Some snapshots taken from one of the experiments are shown in Fig. 7.3.

7.3.1 Influence of Neighborhood Size

The results of the robot experiments are shown in Fig. 7.4. Figure 7.4a reports the behavior of the proportion $(D_a + E_a)/N$ of robots with opinion a over time for the two cases: $G_{max} = 5$ and $G_{max} = 25$. Qualitatively, we observe that the maximum allowed neighborhood size influences the speed of the decision-making process. To determine whether the observed difference in speed was statistically significant, we fitted a generalized linear mixed model (GLMM) (Bolker et al. 2009) with binomial response, where we considered time as a continuous covariate, G_{max} as a fixed factor, and G_{max} nested into the run number as a random factor. In this model, we also included explicitly the interaction of G_{max} with time as an additional fixed factor, which turned out to be significant (p-value = 0.047). The presence of a significant interaction confirms our qualitative observation: the curves representing the predicted proportion of robots with opinion a as a function of time for the two settings ($G_{max} = 5$ and $G_{max} = 25$) do not grow at the same rate. The one with $G_{max} = 25$ grows faster than the one with $G_{max} = 5$. These two curves (dashed lines) are shown in Fig. 7.4b, together with the confidence intervals (shaded areas) predicted by the GLMM analysis. As we can see, the system reaches a 90% majority for opinion a faster with $G_{max} = 25$ than with $G_{max} = 5$: with 95% confidence, the system

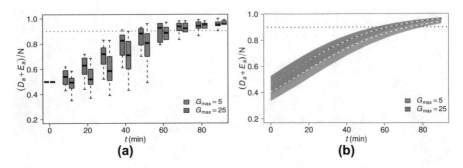

Fig. 7.4 Illustration of the results of the robot experiments and their statistical analysis. **a** shows the distributions of the proportion of robots with opinion a over time. **b** shows the median and the confidence intervals predicted by the GLMM analysis. The *horizontal dotted line* represents a majority of 90% of the swarm favoring opinion a

converges to 90% majority for opinion a in the interval between $t \approx 55$ and $t \approx 69$ for $G_{max} = 25$, and in the interval between $t \approx 66$ and $t \approx 80$ for $G_{max} = 5$.

In both parameter settings, after 90 min of execution the swarm always reached a state where the broad majority of robots favor opinion a, but this majority almost never coincided with 100% consensus. We identified robot failure as a possible cause of this result: robots occasionally experienced battery failures or stuck motors, or switched to stand-by due to short circuits caused by collisions with other robots (0.7 robots per experimental run). Additionally, some robots experienced serious motion difficulties due to poor motor calibration and they were unable to reach target areas (i.e., nest, sites); thus they were prevented from changing opinion. Despite these failures, the DMMD strategy proved to be very robust by allowing the swarm to always reach a correct collective decision.

7.3.2 Group Size and Exploration Time

We performed additional robot experiments to estimate the values of the average group size G in the two settings and that of the time σ^{-1} necessary to explore a site. In the next section, we will use this information to compare the results of the robot experiments with the predictions of the mathematical models described in Chap. 6. Each Kilobot records internally its series of exploration times and that of the number of neighbors at decision time. After an entire experiment, the acquired data are downloaded from the robots using a wired connection. We had to limit the number of experiments for data acquisition because it is a very time-consuming process. Specifically, we performed four runs: two runs to measure the actual average group size in the two settings ($G_{max} = 5$ and $G_{max} = 25$); and two runs to measure the average time required to complete the exploration of a site, again in the two settings.

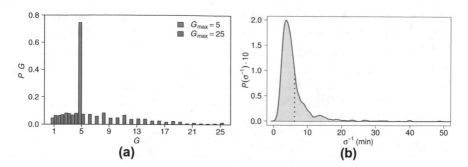

Fig. 7.5 The figure shows the results of the statistical analysis performed during the additional robot experiments. **a** shows the probability mass function of the group size G for the two parameter settings $G_{max} = 5$ (*green*) and $G_{max} = 25$ (purple). **b** shows the probability density function of the time σ^{-1} necessary for a robot to explore a site (the *dotted vertical line* gives the average value of σ^{-1})

Figure 7.5a shows the probability mass function $P(G)$ of the neighborhood size estimated from a single experimental run for each setting. When $G_{max} = 25$ (652 samples), the average group size estimated was 8.57, while it was 4.4 for $G_{max} = 5$ (682 samples). We have therefore a difference of almost a factor of two between the two averages. We performed a two-sample Kolmogorov–Smirnov test to verify that the exploration time has the same probability density function for $G_{max} = 5$ (504 samples) and $G_{max} = 25$ (602 samples). The null hypothesis that the two samples come from the same distribution could not be rejected (p-value = 0.5364), which supports our original conclusion that data sets are consistent with each other. We therefore merged the two data sets to improve our estimate of the exploration time. Figure 7.5b shows the probability density function $P(\sigma^{-1})$ of the time σ^{-1} a robot spends to complete the exploration of a site (we recall that σ is a rate, see Sect. 6.2). The average exploration time is 6.072 min (dotted line).

7.3.3 Comparison with Macroscopic Models

In order to validate our design methodology, we perform a comparison of the robot experiments with the macroscopic mathematical models developed in the previous chapter. Such a comparison is aimed both at validating the predictions of our mathematical models and at gathering additional knowledge about the actual dynamics of the swarm. We initially compare the predictions of the ODE model introduced in Sect. 6.2 with the results of the robot experiments. Successively, we perform the same comparison using instead the predictions of the chemical reaction network introduced in Sect. 6.3. Table 7.2 lists all the parameters used in the ODE model and in the chemical reaction network. We set the group size in both models by rounding the average group size obtained in the robot experiments (cf. Fig. 7.5a). This was 8.57

Table 7.2 Parameters of the ODE model and the chemical reaction network. DP is a design parameter; RE is parameter estimated from robot experiments; PP is a problem parameter

Parameter	Value	Type
Quality of site a	$\rho_a = 1.0$	PP
Quality of site b	$\rho_b = 0.5$	PP
Maximum group size (robots)	$G_{\max} \in \{5, 25\}$	DP
Mean group size (ODEs)	$G \in \{5, 9\}$	RE
Exploration time	$\sigma^{-1} = 6.072\,\text{min}$	RE
Dissemination time	$g = 8.4\,\text{min}$	DP

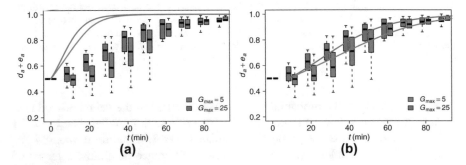

Fig. 7.6 Illustration of the comparison between robot experiments (box-plots) and predictions of the ODE model defined in Chap. 6 (lines), respectively, for $G_{\max} = 5$ (*green*) and $G_{\max} = 25$ (*purple*). **a** shows the comparison of robot experiments with the predictions of the ODE model using the parameters estimated in Sect. 7.3.2. **b** shows the same comparison but the predictions of the ODE model are scaled in time according to $t' = 3t + g$. Parameters: $\sigma = 6.072\,\text{min}$, $g = 8.4\,\text{min}$, $\rho_a = 1$, $\rho_b = 0.5$, $G_{\max} \in \{5, 25\}$, $G \in \{5, 9\}$

when $G_{\max} = 25$ and 4.4 when $G_{\max} = 5$. We therefore set $G = 5$ and $G = 9$ in the two cases. The value of the unbiased dissemination time g was set to $g = 8.4\,\text{min}$. The mean duration of the exploration state (i.e., the inverse of the rate σ at which robots transit from the exploration state to the dissemination state) was estimated from data and equals $\sigma^{-1} = 6.072\,\text{min}$ (cf. Fig. 7.5b).

The comparison between the system of ODEs and the robot experiments is shown in Fig. 7.6a. As we can see, the trajectories predicted by the model (solid lines) have the same shape but do not match those obtained in the robot experiments (box-plots). Specifically, the ODE model appears to be shifted in time and to evolve at a higher speed. Indeed, the fitting improves if we apply the following time rescaling: $t' = 3t + g$, see Fig. 7.6b. This result suggests that robot experiments are approximately 3 times slower than the dynamics predicted by the ODE model, and shifted by a factor g. The offset g is easily explained by assumption *ii.* of the ODE model: initially, the Kilobots do not have a correct estimate of the quality of the two sites but begin the execution with $\rho_a = \rho_b = 1$ (in contrast to assumption *ii.*). Before having a correct quality estimate, the robots have to do an initial exploration of the

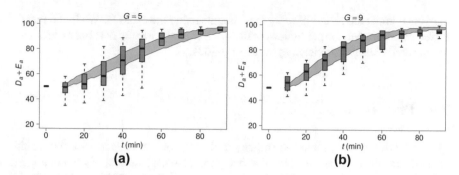

Fig. 7.7 Illustration of the comparison between robot experiments (box-plots) and the predictions of the chemical reaction network in Eqs. (6.7–6.10) approximated by the Gillespie algorithm (shaded areas). The shaded areas correspond to a confidence region computed using the 25th and the 75th percentiles of 1000 independent executions of Algorithm 6.1 with time rescaled according to $t' = 3t + g$. **a** shows the results for the robot scenario with $G_{max} = 5$ and **b** shows the results for $G_{max} = 25$. Parameters: $N = 100$, $\sigma = 6.072\,min$, $g = 8.4\,min$, $\rho_a = 1$, $\rho_b = 0.5$

sites, for which they need to wait on average g minutes. In addition, we conjecture that spatial interference among robots might have caused a partial violation of the well-mixed assumption of the model, which caused a slowdown by a factor of 3. Note that the time rescaling $t' = 3t + g$ has been derived manually without making use of tuning algorithms, thus favoring the simplicity and generality of the resulting explanation over the accuracy of model fitting. Despite this, we obtained correct, qualitative predictions from the ODE model with respect to the asymptotic dynamics of robot experiments. Additionally, we validated the predictions of the ODE model by extending the GLMM model presented in Sect. 7.3.1. We included in the GLMM model the source originating the data as a fixed factor (i.e., robot experiments or ODE model). We then verified that this factor is not statistically significant which means that predictions of the ODE model are not significantly different from the results of the robot experiments (p-value $= 0.436$).

We compare the predictions of the chemical reaction network with the results of the robot experiments. The results of this comparison are shown in Fig. 7.7 as a function of time. The shaded areas provide the region between the 25th and 75th percentile predicted by the model and the box-plots give the outcome of robot experiments. The results of the Gillespie algorithm depicted in the figure are obtained using the same time rescaling $t' = 3t + g$ used for the ODE model. Remarkably, the predictions of the chemical reaction network fit the robot experiments very well for both group sizes $G = 5$ (Fig. 7.7a) and $G = 9$ (Fig. 7.7b). In contrast to the ODE model, the chemical reaction network can also accurately predict the variance of the collective decision-making process. Both in the data from the robot experiments and in the predictions of the model, the variance is higher for intermediate values of time and lower at the beginning and at the end of the execution of the DMMD strategy. As performed for the ODE model, we tested the predictions of the chemical reaction network against the results of robot experiments using the GLMM. We found that the differences between the predictions of the model and the results of robot experiments are not statistically significant (p-value $= 0.367$).

7.4   Discussion

The collective decision-making strategy and the problem scenario studied in this chapter are inspired by the collective behavior of social insects, such as ants and honeybees (Marshall et al. 2009; Franks et al. 2002; Seeley 2010; Sumpter 2010). Specifically, the scenario was inspired by the site-selection problem often faced by honeybee swarms (Franks et al. 2002; Seeley 2010), and was tackled by a swarm of 100 Kilobots (Rubenstein et al. 2014a). The same and similar robots have been successfully used in swarms sized up to one thousand to complete tasks such as aggregation (Kernbach et al. 2009), collective transport (Rubenstein et al. 2013) and pattern formation (Rubenstein et al. 2014b). However, the site-selection scenario analyzed in this chapter is the first experiment in which a large swarm of robots has tackled a consensus achievement problem.

We have shown that the DMMD strategy can be successfully implemented to let a swarm of 100 Kilobots tackle a binary site-selection problem. We have validated the DMMD strategy by performing more than 20 independent repetitions, equivalent to approximately 35 h of robot experiments. The results of the robot experiments show that:

  i the DMMD strategy has sufficiently low requirements to allow its implementation on robots with very limited perception and actuation capabilities;
 ii it is fast enough to implement a feasible collective decision-making process within the robots' limited energy autonomy;
iii it is robust to individual robot failures characteristic of real hardware.

Additionally, we have shown that both the ODE model and the chemical reaction network model described in Chap. 6 yield qualitatively good predictions of the DMMD strategy with an appropriate rescaling of time. The goodness and generality of both models are a function of the value chosen for the design parameter g and of the quality of the estimates of domain-specific parameters (e.g., σ, G, $\rho_i \in \{a, b\}$). Different problem scenarios are likely to require time rescaling different from $3t + g$. The discrepancies between predictions of the models and the results of the robot experiments could be mitigated by increasing the value of g to improve the mixing of robots in space. This would however increase the experimentation time. Conversely, these discrepancies would be exacerbated by decreasing the dissemination time g up to a point where the swarm would approach a macroscopic state of fragmentation.

References

B.M. Bolker, M.E. Brooks, C.J. Clark, S.W. Geange, J.R. Poulsen, M.H.H. Stevens, J.-S.S. White, Generalized linear mixed models: a practical guide for ecology and evolution. Trends Ecol. Evol. **24**(3), 127–135 (2009)

G. Deffuant, D. Neau, F. Amblard, G. Weisbuch, Mixing beliefs among interacting agents. Adv. Complex Syst. **3**((01n04)), 87–98 (2000)

N.R. Franks, S.C. Pratt, E.B. Mallon, N.F. Britton, D.J.T. Sumpter, Information flow, opinion polling and collective intelligence in house-hunting social insects. Philos. Trans. Royal Soc. B Biol. Sci. **357**(1427), 1567–1583 (2002)

S. Kernbach, R. Thenius, O. Kernbach, T. Schmickl, Re-embodiment of honeybee aggregation behavior in an artificial micro-robotic system. Adapt. Behav. **17**(3), 237–259 (2009)

J.A.R. Marshall, R. Bogacz, A. Dornhaus, R. $\tilde{P}$anqué, T. Kovacs, N.R. Franks, On optimal decision-making in brains and social insect colonies. J. Royal Soc Interface **6**(40), 1065–1074 (2009)

N. Mathews, G. Valentini, A.L. Christensen, R. O'Grady, A. Brutschy, M. Dorigo, Spatially targeted communication in decentralized multirobot systems. Auton. Robots **38**(4), 439–457 (2015)

M. Rubenstein, A. Cabrera, J. Werfel, G. Habibi, J. McLurkin, R. Nagpal, Collective transport of complex objects by simple robots: theory and experiments, ed. By T. Ito, C. Jonker, M. Gini, O. Shehory. *Proceedings of the 12th International Conference on Autonomous Agents and Multiagent Systems, AAMAS 2013*(IFAAMAS, 2013), pp. 47–54

M. Rubenstein, C. Ahler, N. Hoff, A. Cabrera, R. Nagpal, Kilobot: a low cost robot with scalable operations designed for collective behaviors. Robot. Auton. Syst. **62**(7), 966–975 (2014a)

M. Rubenstein, A. Cornejo, R. Nagpal, Programmable self-assembly in a thousand-robot swarm. Science **345**(6198), 795–799 (2014b)

T.D. Seeley, *Honeybee Democracy* (Princeton University Press, Princeton, 2010)

D.J.T. Sumpter, *Collective Animal Behavior* (Princeton University Press, Princeton, 2010)

G. Valentini, E. Ferrante, H. Hamann, M. Dorigo, Collective decision with 100 Kilobots: speed versus accuracy in binary discrimination problems (2015a), http://iridia.ulb.ac.be/supp/IridiaSupp2015-005/. Supplementary material, Accessed 24 April 2016

G. Valentini, H. Hamann, M. Dorigo, Self-organized collective decisions in a robot swarm, in *Proceedings of the 29th AAAI Conference on Artificial Intelligence, AI Video Competition* (AAAI Press, 2015b), http://youtu.be/5lz_HnOLBW4. Accessed 24 April 2016

G. Valentini, E. Ferrante, H. Hamann, M. Dorigo, Collective decision with 100 Kilobots: Speed versus accuracy in binary discrimination problems. Auton. Agent. Multi-Agent Syst. **30**(3), 553–580 (2016)

Chapter 8
A Robot Experiment in Collective Perception

In addition to its accuracy and to the time it takes to make a decision, the success of a collective decision-making strategy can be measured by the extent at which it can be generalized across different problem scenarios. Generality of a strategy allows the designer to reuse the existing high-level control logic in different problem scenarios and, while doing so, to focus only on the implementation of domain-specific, low-level control routines (e.g., motion patterns). In this chapter, we support the generality of our design methodology by considering a novel problem scenario: *collective perception*. The collective perception scenario requires a swarm of robots to explore an environment and evaluate the abundance of certain features that are scattered therein (e.g., the availability of precious metals or other minerals) with the objective to determine which feature is the most frequent. We provide domain-specific implementations of the Direct Modulation of Voter-based Decisions strategy and the Direct Modulation of Majority-based Decisions strategy tailored for the e-puck robot. We also consider a third strategy, that we called *Direct Comparison*, with the aim to better highlight the advantages of self-organized approaches. Finally, we use a swarm of e-pucks and study the performance of each strategy over two different problem setups representing a simple and a difficult decision-making problem.

8.1 Robotic Platform and Experimental Setup

We performed experiments using the e-puck robotic platform (Mondada et al. 2009). The e-puck, shown in Fig. 8.1a, is a popular robotic platform within the community of swarm robotics and has been been adopted in a large number of experimental studies. It is a commercially available robot designed for research and education with a diameter of 7 cm and a battery autonomy of up to 45 min. The e-puck is a wheeled robot that can move with a maximum speed of 16 cm/s. In its basic

© Springer International Publishing AG 2017

G. Valentini, *Achieving Consensus in Robot Swarms*, Studies in Computational Intelligence 706, DOI 10.1007/978-3-319-53609-5_8

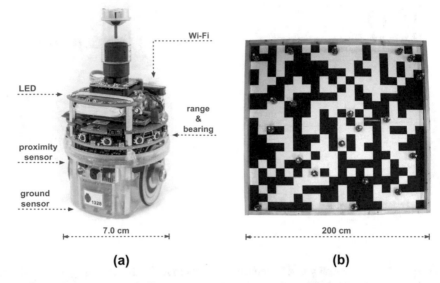

Fig. 8.1 Illustration of the robotic platform and the experimental setup. Figure **a** shows the e-puck robot highlighting, in clock-wise order, the position of the Wi-Fi board, the range and bearing board, the ground sensor, the proximity sensors, and the RGB LEDs. Figure **b** shows a *top-view* picture of the arena used for the collective perception scenario with a swarm of $N = 20$ e-pucks, 10 with opinion a (*red* LEDs) and 10 with opinion b (*blue* LEDs). The *color pattern* drawn on the surface of the arena represents a difficult problem scenario with 52% of the surface colored in *black* and 48% colored in *white*

configuration, the robot is equipped with RGB LEDs, a low-resolution camera, an accelerometer, a sound sensor, and 8 proximity sensors. Figure 8.1a shows the e-puck configuration used in our experiments where the robot is extended with the range and bearing IR communication module (Gutiérrez et al. 2009), the ground sensor, the Overo Gumstick module, and the omnidirectional turret.[1] In our experiments, the robots use the range and bearing module to share their information locally with their neighbors (e.g., internal state, quality estimate). This module consists of 12 IR transceivers positioned around the circumference of the robot that allow it to send and receive messages up to a distance of approximately 70 cm. Additionally, the e-puck mounts 8 IR proximity sensors that are used to detect the presence and measure the distance of nearby obstacles. The e-puck has 3 ground sensors that allow it to measure the color of the surface in gray-scale values. Finally, the Overo Gumstick module provides the e-puck with the capabilities to run Linux and with a Wi-Fi connection. This latter feature is exploited during our experiments to collect statistics about the collective decision-making process.

For both robot experiments (see Fig. 8.1b) and physics-based simulations (see Fig. 8.2), we consider a collective perception scenario characterized by an environ-

---

[1]The omnidirectional camera mounted in the e-puck turret is not used by robots in the collective perception scenario.

(a)

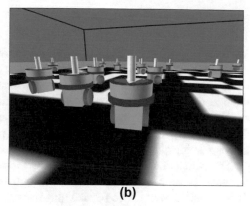
(b)

Fig. 8.2 Illustration of the physics-based simulations of the collective perception scenario implemented using the ARGoS simulator (Pinciroli et al. 2012). Figure **a**, **b** show, respectively, a *top* and a *side view* of the simulated environment with a swarm of $N = 20$ e-pucks

ment with $n = 2$ features. The robots are positioned in a square arena with a total area of 200×200 cm^2. As in Chap. 7, the environment is approximately three orders of magnitude larger than a single robot footprint. It is bounded by four walls that can be detected by the proximity sensors of the robots. The surface of the environment is characterized by a grid consisting of 10×10 cm^2 cells. The color of each cell is used as an abstraction to represent a particular feature of the environment. Robots always have an opinion about which feature they currently believe to be the most frequent. In particular, the color black represents the feature of the environment associated to opinion a while the color white represents the feature of the environment associated to opinion b. Without loss of generality, we always have the black feature as the most frequent in the environment and, as a consequence, the goal of the swarm is to make a collective decision favoring opinion a. Each robot of the swarm uses its LEDs to show its current opinion. LEDs are lighted up in red when the robot favors opinion a and in blue when the robot favors opinion b. The robots use their ground sensors to perceive the brightness of the underlying surface, determine its color, and estimate the quality of the corresponding option.

8.2 Robot Control Algorithm

In this section, we describe the three collective decision-making strategies used in our performance comparison. All three strategies rely on common low-level control routines (i.e., random walk, obstacle avoidance, and quality estimation) that are described in Sect. 8.2.1. In Sect. 8.2.2, we describe the implementation of the DMMD strategy and the DMVD strategy for the collective perception scenario. Section 8.2.3 provides instead the description of the Direct Comparison (DC) strategy.

8.2.1 Low-Level Motion Routines

We implemented a random walk routine as follows. A robot performing random walk alternates between straight motion and rotation on the spot. The robot moves straight for a random period of time with a mean duration of 40 s that is sampled from an exponential distribution. After this period of time, the robot turns on the spot for a random period of time that is uniformly distributed between 0 and 4.5 s. The turning direction is also chosen randomly. With equal probability, the robot turns clockwise or counterclockwise. Once turning is completed, the robot resumes straight motion.

The perception by a robot of one or more nearby obstacles (i.e., a wall or a neighboring robot at a distance less than approximately 30 cm) causes the execution of the random walk to be paused and triggers the obstacle avoidance routine. We implemented the obstacle avoidance routine as follows. The robot uses its proximity sensors to detect the distance and the bearing of each perceived obstacle. It then uses this information to compute a new direction of motion that is opposite to the obstacles. Depending on the computed direction, the robot turns on the spot either clockwise or counterclockwise until its orientation corresponds to the computed one. Then, the robot resumes its random walk.

We implemented the following quality estimation routine to let a robot estimate the quality ρ_i of the feature associated to its opinion $i \in \{a, b\}$. When executing the quality estimation routine, the robot uses its ground sensors to sample the color of the surface while moving randomly in the environment. During the entire execution of the quality estimation routine, the robot keeps track of the amount of time τ_i during which it perceived the color associated to its current opinion i. Finally, the robot computes a quality estimate $\hat{\rho}_i$ which is the ratio of τ_i to the overall duration of the quality estimation routine.

8.2.2 DMMD and DMVD Strategies Implementation

In the exploration states E_i, $i \in \{a, b\}$, a robot with opinion i explores the environment by performing the random walk routine and, when necessary, the obstacle avoidance routine. Meanwhile, the robot samples the environment locally and estimates the option quality ρ_i, by executing the quality estimation routine. The duration of the exploration state is random and exponentially distributed with a mean duration of σ^{-1} s (cf. Chap. 6). After this period of time is elapsed, the robot switches to the dissemination state D_i.

In the dissemination states D_i, $i \in \{a, b\}$, a robot with opinion i broadcasts its opinion locally to its neighbors using its range and bearing module. Meanwhile, the robot performs the same random walk and obstacle avoidance routines as in the exploration states. The aim of this motion pattern, however, is not to explore the environment but to mix the positions of robots of different opinions in the environment which eases the collective decision-making process. The robot uses its current quality estimate $\hat{\rho}_i$ to amplify or inhibit the duration of the dissemination state D_i in a way that this duration is proportional to the opinion quality. To do so, the duration of the dissemination state is exponentially distributed with mean $\hat{\rho}_i g$ s, where g is a design parameter that defines the unbiased dissemination time. Finally, the robot collects the opinions broadcast by its neighbors and applies the individual decision-making mechanism (either the majority rule in the DMMD strategy or the voter model in the DMVD strategy) to determine its new opinion $j \in \{a, b\}$. Then, the robot switches to the exploration state E_j to collect a new estimate $\hat{\rho}_j$ of the option quality.

8.2.3 Direct Comparison of Option Quality

We define a third decision-making strategy, the Direct Comparison of option quality, by using the same PFSM of the DMMD and DMVD strategies but letting robots compare their quality estimates directly to modify their opinion. When executing the DC strategy, robots alternate periods of exploration to periods of dissemination. In contrast to the DMVD and DMMD strategies, the DC strategy does not make use of a mechanism for the modulation of positive feedback and the mean duration of the dissemination state D_i, $i \in \{a, b\}$, is g, independently of the option quality ρ_i. During the dissemination period, the robot also broadcasts its current quality estimate $\hat{\rho}_i$ in addition to its opinion i. This additional information is used by robots to modify their opinions. At the end of the dissemination state, a robot with opinion i compares its opinion with that of a random neighbor with opinion $j \in \{a, b\}$. If the neighbor's estimate $\hat{\rho}_j$ is greater than the considered robot's estimate $\hat{\rho}_i$, then the robot modifies its current opinion to j. Next, the robot switches to the exploration state E_j which is implemented identically to that of the DMMD and DMVD strategies.

Our aim is to use the DC strategy to show the benefits of a self-organized approach. Indeed, as observed in natural systems (Edwards and Pratt 2009; Visscher and Camazine 1999), the ability of the DMMD and DMVD strategies to discriminate different options is based on the self-organized processing of a multitude of individual quality estimates by the swarm. These quality estimates are processed by modulating positive feedback in combination with an individual decision-making mechanism that operates on opinions of neighbors only. In contrast, a swarm executing the DC strategy relies on the capabilities of individual robots to correctly discriminate the different options by their quality.

8.3 Experiments

We consider the collective perception scenario described in Sect. 8.1 and perform experiments using the DMVD, DMMD, and DC strategies. Our primary objective during these experiments is to compare the performance of the three considered collective decision-making strategies in terms of the speed and the accuracy of the resulting collective decisions (Franks et al. 2003; Passino and Seeley 2006). As done in the previous chapters, we measure the strategies' speed using the average time T_N necessary for a swarm to reach consensus on any opinion, while we use the exit probability E_N, computed as the proportion of experimental runs that converge to consensus on opinion a, as a measure of the strategies' accuracy. By doing so, we aim at investigating if the additional information used by the DC strategy (i.e., the neighbors' estimates of the option quality) provides any benefits with respect to the more canonical approach underlying the DMVD and the DMMD strategies. In the following, we first perform experiments using a swarm of $N = 20$ e-pucks in two different setups of the environment classification scenario representing a simple and a more difficult decision-making problem.[2] Then, we deepen our experimental analysis by means of physics-based simulations implemented using the ARGoS simulator (Pinciroli et al. 2012). Physics-based simulations allows us to study the behavior of the DMVD, DMMD, and DC strategies over a wider space of parameters configurations including the initial conditions, the problem difficulty, and the swarm size. In both robot experiments and physics-based simulations, we set robots to start the execution of their controllers in the exploration state. Additionally, we set the mean duration of the exploration state to $\sigma^{-1} = 10\,\mathrm{s}$ and the unbiased duration of the dissemination state to $g = 10\,\mathrm{s}$.

8.3.1   Robot Experiments

We considered two different experimental setups for the collective perception problem. The first setup represents a simple decision-making problem where the proportion of feature a (i.e., color black) in the environment is approximately twice that of feature b (i.e., color white). Specifically, the surface of the environment is $\rho_a = 66\%$ black and $\rho_b = 34\%$ white and the problem difficulty is defined by the normalized option qualities $\rho_a^\star = 1$ and $\rho_b^\star = \rho_b/\rho_a = 0.515$. The second setup consists of a more difficult collective perception problem where the surface of the environment is $\rho_a = 52\%$ black and $\rho_b = 48\%$ white (i.e., $\rho_a^\star = 1$ and $\rho_b^\star = 0.923$). For each combination of problem setup and collective decision-making strategy, we performed 15 repetitions of the robot experiment (i.e., a total of 90 repetitions). In all experiments, the swarm is initially unbiased with 10 robots in state E_a and 10 robots in state E_b.

Figure 8.3 shows the results of the robot experiments for the simple collective perception scenario for the DMMD strategy (top), the DMVD strategy (middle), and the DC strategy (bottom). The box-plots provide the evolution over time of the number $D_a + E_a$ of robots with opinion a: white box-plots for the repetitions converging to consensus on opinion a and gray box-plots for the repetitions converging to consensus on opinion b. The vertical lines indicate the average time T_N to

---

[2] All robot experiments have been recorded and videos can be found online in the supplementary material (Valentini et al. 2016b) of the article (Valentini et al. 2016a).

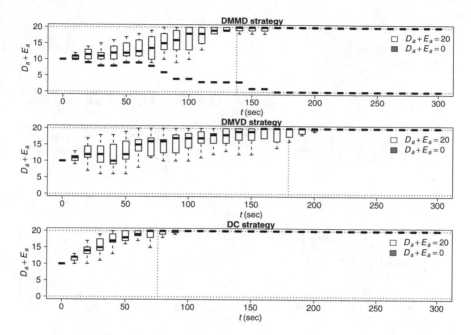

Fig. 8.3 Illustration of the results of the robot experiments for the simple experimental setup. The figure shows the evolution over time of the number of robots with opinion a (i.e., $D_a + E_a$) for the DMMD strategy (*top*), the DMVD strategy (*middle*), and the DC strategy (*bottom*). *White* and *gray* box-plots shows, respectively, the distribution of the experimental runs converging to consensus on opinion a and consensus on opinion b. The *gray* box-plots are not plotted in the case in which all runs converged on opinion a. The vertical lines show the average time necessary to reach consensus on any opinion. Parameters: $\rho_a^\star = 1$, $\rho_b^\star = 0.515$, $\sigma^{-1} = 10$ s, $g = 10$ s, $N = 20$

reach consensus. When executing the DMMD strategy (see Fig. 8.3, top), the swarm of e-pucks requires on average $T_N = 138$ s to converge on a consensus decision with a standard deviation of 35.5 s; its accuracy is $E_N = 0.933$, i.e., only 1 out of 15 repetitions converges to a wrong consensus on opinion b. In contrast, when executing the DMVD strategy or the DC strategy, the swarm of e-pucks is always able to correctly identify the most frequent feature in the environment (i.e., decision accuracy $E_N = 1.0$). However, the DMVD strategy converges to consensus after $T_N = 179.3$ s with a standard deviation of 108.4 s while the DC strategy is faster and requires only $T_N = 76$ s and has a standard deviation of 35.2 s. In agreement with the results of Chap. 6, we observe that the DMMD strategy is faster but also less accurate than the DMVD strategy. For the simple experimental setup, the DC strategy benefits from using more information; it is faster than both the DMMD strategy and the DMVD strategy and has the same maximum accuracy as the DMVD strategy.

Figure 8.4 shows the results of the robot experiments in the difficult setup of the collective perception scenario in which the normalized option qualities are given by $\rho_a^\star = 1$ and $\rho_b^\star = 0.923$. As for the simple experimental setup, Fig. 8.4 shows the results for the DMMD strategy (top), the DMVD strategy (middle) and the DC

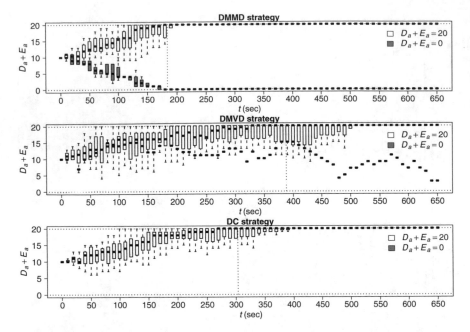

Fig. 8.4 Illustration of the results of the robot experiments for the difficult experimental setup. The figure shows the evolution over time of the number of robots with opinion a (i.e., $D_a + E_a$) for the DMMD strategy (*top*), the DMVD strategy (*middle*), and the DC strategy (*bottom*). *White* and *gray* box-plots shows, respectively, the distribution of the experimental runs converging to consensus on opinion a and consensus on opinion b. The *gray* box-plots are not plotted in the case in which all runs converged on opinion a. The *vertical lines* show the average time necessary to reach consensus on any opinion. Parameters: $\rho_a^\star = 1$, $\rho_b^\star = 0.923$, $\sigma^{-1} = 10$ s, $g = 10$ s, $N = 20$

strategy (bottom). The increased difficulty of the decision-making problem overturns the results obtained in the simple experimental setup. The DMMD strategy based on the majority rule is the fastest strategy in the comparison with an average consensus time of $T_N = 184$ s and a standard deviation of 64.8 s. The DMVD strategy based on the voter model is still the slowest strategy with an average consensus time of $T_N = 387.9$ s and a standard deviation of 291.2 s. The DC strategy has an average consensus time of $T_N = 303.3$ s and a standard deviation of 135.2 s. In contrast, the DMMD strategy has the lowest accuracy, $E_N = 0.667$, reaching consensus on the best option 10 times out of 15 repetitions. The DMVD strategy, with a decision accuracy of $E_N = 0.933$, performs similarly to the DC strategy whose decision accuracy is still maximal, $E_N = 1.0$. For higher difficulty of the collective perception scenario, there is no strategy that outperforms all others in both speed and accuracy.

The communication overhead underlying the DC strategy seems to provide stronger benefits than those of the modulation of positive feedback used by the DMMD and DMVD strategies. However, a cross comparison of the results between the simple and difficult experimental setups reveals an interesting performance trend. The increase in the difficulty of the decision-making problem resulted in a relative little slowdown of the DMMD strategy which is 1.33 times slower when compared to the simple experimental setup; the DMVD strategy is 2.16 times slower; while the DC strategy has a more pronounced slow down of 3.99 times. The DMMD strategy, with an accuracy 28.5% less than the simple setup, is preferable when consensus time is the most critical constraints. The DMVD strategy loses only 6.7% of its accuracy and its consensus time increases much less than that of the DC strategy. This trend suggests that the DMVD strategy could be the choice of reference for the designer when favoring the accuracy of the collective decision.

The results of our robot experiments provide us with useful indications; however, since such experiments are particularly time-consuming, we could collect a limited amount of data (i.e., only 15 independent repetitions for each parameter configuration). As a consequence, the statistics computed from the robot experiments are characterized by a pronounced spread which is shown by the standard deviation of the consensus time. In the next section, we deepen the results of our analysis by means of physics-based simulations.

8.3.2 Physics-Based Simulations

We performed physics-based simulations using the ARGoS simulator (Pinciroli et al. 2012) and compared the performance of the DMMD, DMVD, and DC strategies over a wider region of the parameter space than what we did in the robot experiments. We varied the initial number $E_a(0)$ of robots favoring opinion a, the swarm size N, and the difficulty of the collective perception scenario through the normalized option quality $\rho_b^\star$. For each parameter configuration, we collected data from 1000 independent repetitions of the simulated experiment.

We set $N = 20$ and study the simple and difficult scenarios defined above as a function of the initial number $E_a(0)$ of robots with opinion a (see Fig. 8.5). For the simple scenario, the exit probability E_N of the three strategies corresponds to that obtained in the robot experiments (cf. $E_a(0) = 10$ in Fig. 8.5a). For all strategies, E_N increases with increasing values of the initial number $E_a(0)$ of robots with opinion a; the DC strategy has the highest accuracy while the DMMD strategy has the lowest. However, for all three collective decision-making strategies, the consensus time T_N shown in Fig. 8.5b is considerably shorter than that obtained with robot experiments. Additionally, the DMMD strategy is now the fastest strategy and it outperforms the DC strategy for all initial conditions $E_a(0)$. For the difficult scenario, we observe similar differences in the speed and accuracy of the three decision-making strategies. Both the DMVD and DC strategies are considerably less accurate than in the robot experiments (see Fig. 8.5c). In addition, the decision accuracy of the DMMD strategy decreases more slowly than that of the DMVD and DC strategies when decreasing the value of $E_a(0)$. As for the simple scenario, all strategies converge faster to consensus (see Fig. 8.5d). The DC strategy is the slowest strategy in the comparison.

The results of physics-based simulations reproduce only partially the performance obtained with robot experiments. The observed discrepancies are a result of differ-

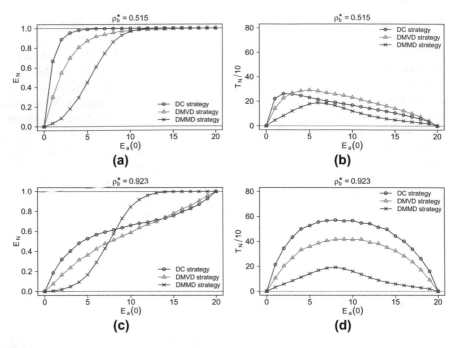

Fig. 8.5 Illustration of the exit probability E_N and of the consensus time T_N as a function of the initial number of robots favoring opinion a. Figure **a**, **b** show, respectively, the exit probability and the consensus time for a simple decision-making problem with $\rho_b^\star = 0.515$. Figure **c**, **d** show the same metrics but for a more difficult decision-making problem where $\rho_b^\star = 0.923$. Parameters: $\rho_a^\star = 1.0$, $\rho_b^\star \in \{0.515, 0.923\}$, $\sigma^{-1} = 10\,\mathrm{s}$, $g = 10\,\mathrm{s}$, $N = 20$

ences in the level of noise between simulation and reality. For example, a few robots used during the experiments have particularly noisy proximity sensors; as a result, they often collide with other robots or with the walls. Additionally, the uneven surface of the experimental arena caused robots to remain temporarily stuck over the same cell resulting into particularly erratic quality estimates. The influence of these factors is exacerbated by the limited number of runs performed with real robots as shown by the large spread of the consensus time characterizing the results in Figs. 8.3 and 8.4. Nonetheless, the physics-based simulations confirm a poor scalability of the DC strategy as previously shown by the robot experiments.

We deepen our comparison by analyzing the speed and the accuracy of the DMMD, DMVD, and DC strategies when varying the swarm size N and the difficulty $\rho_b^\star$ of the collective perception scenario (see Fig. 8.6). The exit probability of all considered strategies decreases for increasing difficulty of the collective perception scenario (see Fig. 8.6a, b). However, the DMVD and DC strategies benefit from the bigger swarm size, $N = 100$, as shown by their higher accuracy. In contrast, the DMMD strategy is affected by the increased swarm size and the resulting collective decisions are less accurate. For swarms of size $N = 20$ and $N = 100$, we observe that the

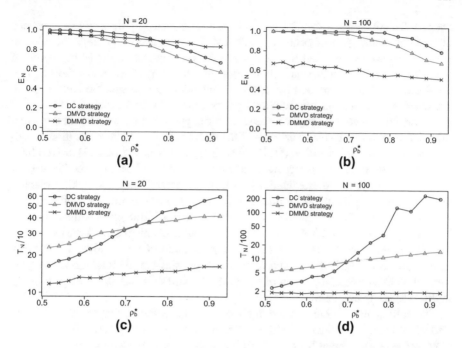

Fig. 8.6 Illustration of the exit probability E_N and the consensus time T_N for decision-making problems of increasing difficulty (i.e., the normalized option quality $\rho_b^\star \to \rho_a^\star$). Figure **a, b** show the exit probability, respectively, for a swarm of $N = 20$ and $N = 100$ e-pucks. Figure **c, d** show the consensus time as a function of $\rho_b^\star$, respectively, for a swarm of $N = 20$ and $N = 100$ simulated e-pucks. Note that the vertical axis is characterized by a logarithmic scale. Parameters: $\rho_a^\star = 1.0$, $\rho_b^\star \in [0.515; 0.923]$, $\sigma^{-1} = 10\,\text{s}$, $g = 10\,\text{s}$, $N \in \{20, 100\}$, $E_a(0) = N/2$, $E_b(0) = N/2$

DC strategy is the strategy that suffers the fastest slowdown of the consensus time as a result of increasing the difficulty of the decision-making problem. This result confirms the trend observed from the analysis of the robot experiments. Additionally, by comparing Fig. 8.6c with 8.6d, we also observe that the consensus time of the DC strategy increases much faster than that of the other strategies when the size of the swarm is increased to $N = 100$; therefore, the DC strategy does not scale with the swarm size. Contrarily, the consensus time of the DMMD and DMVD strategies is only slightly affected by the larger swarm size.

8.4 Discussion

In this chapter, we described a novel collective decision-making scenario, referred to as collective perception, that requires a swarm of robots to explore a certain environment, perceive the presence of certain features, and determine which feature is the most frequent. As for the site-selection scenario considered in the previous chapter, the collective perception scenario is characterized by the complete absence of indirect modulation of positive feedback. The absence of indirect modulation is a consequence of the fact that the entire swarm operates in the same region of the environment (and thus spatiality cannot introduce any bias) and of the fact that the duration of the estimation routine is independent of the specific feature being evaluated at that moment (represented by the color of the arena surface).

Using the collective perception scenario, we supported the generality of the direct modulation of majority-based decision (DMMD) strategy and that of the direct modulation of voter-based decision (DMVD) strategy. DMMD and DMVD are modular strategies that combine a direct modulation mechanism of positive feedback (i.e., modulation of opinion dissemination) with an individual decision-making mechanism (respectively, the majority rule and the voter model). In order to better understand the benefits of these modules, we considered a third strategy, the direct comparison of option quality (DC), that has no modulation mechanism and whose individual decision-making mechanism relies on a larger amount of information (i.e., quality estimates). Using both robot experiments and physics-based simulations, we performed an extensive comparison of the DMMD, DMVD, and DC strategies under realistic working conditions.

Our results are twofold. On the one hand, we have shown that the DMMD and DMVD strategies provided us with off-the-shelf solutions to a collective decision-making scenario different from their original context of site-selection. When applied to the collective perception scenario, these design solutions maintained the same speed and accuracy performance observed for the site-selection scenario and showed a promising level of generality. On the other hand, we have shown that, despite relying on less information, the self-organized approach of the DMMD and DMVD strategies is more robust to noise and individual errors. We have done so by highlighting the scalability problems of the DC strategy for increasing problem difficulty, ρ_b/ρ_a, and/or swarm size, N. When adopted by an individual agent, the direct comparison of options' quality is known to perform poorly for difficult decision-making problems (Sasaki et al. 2013) as well as for problems characterized by many options (Sasaki and Pratt 2012). Our results show that the larger amount of information used by the DC strategy is not beneficial even when it is used in a distributed manner by a collective of agents.

References

S.C. Edwards, S.C. Pratt, Rationality in collective decision-making by ant colonies. Proc. R. Soc. B: Biol. Sci. **276**(1673), 3655–3661 (2009)

N.R. Franks, A. Dornhaus, J.P. Fitzsimmons, M. Stevens, Speed versus accuracy in collective decision making. Proc. R. Soc. B: Biol. Sci. **270**, 2457–2463 (2003)

Á. Gutiérrez, A. Campo, M. Dorigo, J. Donate, F. Monasterio-Huelin, L. Magdalena, Open e-puck range & bearing miniaturized board for local communication in swarm robotics, in *IEEE International Conference on Robotics and Automation, ICRA '09*, (2009), pp. 3111–3116

F. Mondada, M. Bonani, X. Raemy, J. Pugh, C. Cianci, A. Klaptocz, S. Magnenat, J.-C. Zufferey, D. Floreano, A. Martinoli, The e-puck, a robot designed for education in engineering, in *Proceedings of the 9th Conference on Autonomous Robot Systems and Competitions*, vol. 1, ed. by P.J.S. Gonçalves, P.J.D. Torres, C.M.O. Alves (IPCB, Instituto Politécnico de Castelo Branco, 2009), pp. 59–65

K.M. Passino, T.D. Seeley, Modeling and analysis of nest-site selection by honeybee swarms: the speed and accuracy trade-off. Behav. Ecol. Sociobiol. **59**(3), 427–442 (2006)

C. Pinciroli, V. Trianni, R. O'Grady, G. Pini, A. Brutschy, M. Brambilla, N. Mathews, E. Ferrante, G. Di Caro, F. Ducatelle, M. Birattari, L.M. Gambardella, M. Dorigo, ARGoS: a modular, parallel, multi-engine simulator for multi-robot systems. Swarm Intell. **6**(4), 271–295 (2012)

T. Sasaki, S.C. Pratt, Groups have a larger cognitive capacity than individuals. Curr. Biol. **22**(19), R827–R829 (2012)

T. Sasaki, B. Granovskiy, R.P. Mann, D.J.T. Sumpter, S.C. Pratt, Ant colonies outperform individuals when a sensory discrimination task is difficult but not when it is easy. Proc. Natl. Acad. Sci. **110**(34), 13769–13773 (2013)

G. Valentini, D. Brambilla, H. Hamann, M. Dorigo, Collective perception of environmental features in a robot swarm, in *Swarm Intelligence*, volume 9882 of *LNCS* ed. By M. Dorigo, M. Birattari, X. Li, M. López-Ibáñez, K. Ohkura, C. Pinciroli, T. Stützle (Springer, 2016a), pp. 65–76

G. Valentini, D. Brambilla, H. Hamann, M. Dorigo, Collective perception of environmental features in a robot swarm (2016b). http://iridia.ulb.ac.be/supp/IridiaSupp2016-002/. Supplementary material, accessed 24 Apr 2016

P.K. Visscher, S. Camazine, Collective decisions and cognition in bees. Nature **397**(6718), 400 (1999)

Part IV
Discussion and Annexes

Chapter 9
Conclusions

We conclude by summarizing the primary principles of discrete consensus achievement in robot swarms as introduced in this book. We review the motivations of our work and highlight the scientific contributions provided in each chapter. Finally, we provide the reader with the take-home message of this book and discuss possible venues for future developments.

9.1 Motivations and Scientific Contributions

Consensus achievement refers to the phenomenon whereby agents in a collective gather information from the environment, pool this information with each other, and process it to obtain a collective decision addressing a certain cognitive problem. The ability to undertake a collective decision-making process and to achieve consensus is a cognitive skill of paramount importance in robot swarms. It allows the swarm to function as a compact information processing entity that overcomes the cognitive constraints that affect individual agents. Throughout a collective decision-making process, the agents of a swarm sample information from the environment in parallel extending their reach and gathering a greater amount of knowledge than is possible for a single individual. This information, which may exceed the processing capacity of single agents, is processed by means of self-organized processes that result from the execution of interaction rules sufficiently simple for agents with limited capabilities. In addition to non-recurring cognitive problems, consensus achievement is often required to coordinate the multitude of agents composing robot swarms. For example, complex application scenarios can be favorably decomposed into a series of tasks that the swarm should perform one after the other; in this situation, coordination may require to achieve consensus on when to start working on the next task of the sequence.

© Springer International Publishing AG 2017

G. Valentini, *Achieving Consensus in Robot Swarms*, Studies in Computational Intelligence 706, DOI 10.1007/978-3-319-53609-5_9

Despite its relevance, the study of consensus achievement in robot swarm has received little attention from the swarm robotics community. In particular, the research described in this book is motivated by the lack of a comprehensive theoretical framework for the design and analysis of collective decision-making strategies for discrete consensus achievement problems. That is, decision-making problems that require a swarm of agents to achieve consensus over which option of a finite set of alternatives is the solution that is most advantageous for the swarm. With the aim of advancing our understanding of discrete consensus achievement, we put forward a principled methodology to design collective decision-making strategies for robot swarms. The methodology we propose builds on the understanding of the fundamental mechanisms underlying collective decision making to dictate a modular and model-driven perspective on the design of collective decision-making strategies.

Our approach is based on the abstraction of the cognitive problem from the specific application scenario as well as on the separation of the problem from the collective decision-making strategy providing its solution. Using this approach we were able to focus separately on the different processes that characterize the decision-making process and to isolate and identify their generating mechanisms. In the following, we summarize the contributions that resulted from this divide and rule approach.

In Chap. 2, we provided the first in-depth review of the literature of discrete consensus achievement from a swarm robotics perspective. We formalized the logic and structure of the best-of-n problem and showed how different combinations of internal preference factors and environmental bias factors determine specific variants of this problem. By adopting this perspective, we reviewed studies in swarm robotics that deal with discrete consensus achievement organizing them according to their design approach: opinion-based approaches, ad hoc approaches, and automatic approaches. In doing so, we showed that only opinion-based and automatic approaches have the sufficient generality to design collective decision-making strategies across different application scenarios; and that, among these two design approaches, only opinion-based approaches are amenable for the derivation of predictive macroscopic mathematical models. The results in Chap. 2 support the use of an opinion-based approach as is the case for the modular design methodology proposed in this book.

In Chap. 3, we detailed a modular and model-driven design methodology to define collective decision-making strategies for the best-of-n problem. We identified four mechanisms that are necessary to achieve consensus: option exploration, to gather information on the quality of an option; opinion dissemination, to pool the gathered information with other members of the swarm; modulation of positive feedback, to bias the decision-making process in favor of the best option; and individual decision-making mechanism, to allow agents to change opinion about the best option of the decision-making problem. We showed how modulation of positive feedback can be distinguished in direct modulation, when it results from agents adjusting their behavior as a function of internal preference factors, and indirect modulation, when environmental bias factors influence the swarm dynamics. By building on this understanding, we have proposed a general and modular structure of a collective decision-making strategy that implements all fundamental mechanisms and have provided guidelines as well as constraints to design specific modules. This approach allowed us to define a generic model of a collective decision-making strategy that can be instantiated by the designer to study a specific combination of modules defining a strategy or to readily compare different design alternatives implementing specific modules.

In Chaps. 4, 5 and 6, we illustrated the application of our modular design methodology. In Chap. 4, we provided an interpretation of the majority rule with differential latency (Montes de Oca et al. 2011) according to our framework and showed how this strategy combines an indirect mechanism for the modulation of positive feedback to the majority rule functioning as the individual decision-making mechanism (i.e., *Indirect Modulation of Majority-based Decisions* strategy, IMMD). We analyzed the IMMD strategy by means of an absorbing Markov chain model and provided novel insights on the variance of consensus time. In Chaps. 5 and 6, we instantiated our design methodology to define and analyze two novel strategies: the *Direct Modulation of Voter-based Decisions* strategy (DMVD) and the *Direct Modulation of Majority-based Decisions* strategy (DMMD). Both the DMVD and DMMD strategies make use of a mechanism for the direct modulation of positive feedback in order to bias the decision-making process as a function of one or more internal preference factors. However, the two strategies differ in the choice of the individual decision-making mechanism: the DMVD strategy is based on the voter model and the DMMD strategy is based on the majority rule. We studied the asymptotic properties of both strategies by means of ordinary differential equations and finite-size effects using chemical reaction networks. In doing so, we showed that the DMVD and DMMD strategies provide different design compromises in terms of the speed and accuracy of the collective decision. In particular, we showed that the DMVD strategy is characterized by a pronounced accuracy of the resulting collective decision but that requires long time to achieve consensus. In contrast to the DMVD strategy, the DMMD strategy is characterized by lower decision accuracy and allows the swarm to make collective decision in shorter time.

In Chap. 7, we validated our design methodology experimentally by implementing the DMMD strategy in a site-selection scenario using a swarm of 100 Kilobots. In this application scenario, the robots of the swarm need to choose between two sites based on an internal preference factor determining the quality of each site. We studied the speed versus accuracy trade-off that characterizes the DMMD strategy as a function of the neighborhood size. We showed that the DMMD strategy can be successfully implemented using robots with limited perception and actuation capabilities, that is sufficiently fast to achieve consensus within the robots' energy autonomy, and that is robust to failures of individual robots. Additionally, we showed that our modeling methodology provides qualitatively good predictions of the dynamics of a swarm executing the DMMD strategy. We have done so by comparing the results of robot experiments with the predictions of both the deterministic and the stochastic mathematical models introduced in Chap. 6.

In Chap. 8, we further showcased our design methodology by performing a series of experiments using a swarm of 20 e-pucks (using both physics-based simulations and real-robot experiments). We proposed a novel application scenario—the *collective perception* scenario—that requires a swarm of robots to determine the most frequent feature in the environment. Using the collective perception scenario, we supported the generality of the DMVD and DMMD strategies. We achieve this objective by showing how these strategies provided off-the-shelf solutions to an application scenario different from their original one (i.e., site selection) while maintaining the same performance in terms of the speed and accuracy of the collective decision. In Chap. 8, we also studied the *Direct Comparison of options' quality* (DC), a collective decision-making strategy that has no modulation mechanism and in which agents change opinion only when perceiving a neighbor with a better quality estimate. By comparing the performance of the DMVD and DMMD strategies to that of the DC strategy, we showed how indirect information processing resulting from the combination of modulation of positive feedback with a simple individual decision-making mechanism allows the DMVD and DMMD strategies to be robust to noise and individual errors.

The research contributions provided in this book represent a relevant step forward towards the definition of a formal theoretical framework for the design and analysis of collective decision making in robot swarms. However, our contribution in this direction is far from being complete and leaves space for future developments and enhancements.

9.2   Take-Home Message and Future Directions of Research

In this book we have seen how collective decision-making strategies for the best-of-n problem can be conveniently designed by leveraging on the concept of modularity. Four modules are necessary to design a strategy: option exploration, opinion dissemination, modulation of positive feedback, and individual decision-making mechanism. The take-home message for the reader of this book is articulated in four points:

(i) The option exploration and opinion dissemination modules are required to gather and pool information about the options of the best-of-n problem. The implementation of these modules largely depends on the specific application scenario. However, the designer needs only to implement these modules to reuse an existing strategy in a different scenario. This modular approach promotes the generality of the designed strategies.

(ii) Modulation of positive feedback is required to drive the swarm towards consensus for the best option. It can be performed directly by the agents of the swarm or indirectly through the interaction between the swarm and the environment. Direct modulation is entirely under the control of the designer while indirect modulation depends on the application scenario. Depending on the target scenario, the designer might rely only on indirect modulation to design simple strategies that minimize the costs of a collective decision; she/he might include direct modulation to further increase the performance of the strategy or to compensate for the negative influence of the environment.

(iii) The individual decision-making mechanism is needed for the agents to process the information gathered by the swarm and to change their opinion. This module allows the designer to shape the performance of a strategy in terms of the speed and the accuracy of the resulting collective decision. The designer might favor the speed of the collective decision at the expense of its accuracy, for example, using the majority rule, or vice versa, she/he might prioritize accuracy over speed, for example, using the voter model.

(iv) Modulation of positive feedback and individual decision-making mechanism can be designed independently of other modules and then combined in different configurations to create different strategies. This modular approach eases the definition of mathematical models that support the design process and allow for the creation of a catalog of modules with known properties.

These four points summarize the guidelines necessary to design new modules of a collective decision-making strategy beyond those investigated in this book. The model-driven design methodology introduced in this book provides the means to model and to study the dynamics of newer strategies supporting the development of a theory of collective decisions.

As we showed in Chap. 2, the combination of internal preference and environmental bias factors determines the particular variant of the best-of-n problem. The strategies described in this book directly apply to all possible variants. However, when environmental bias factors affect negatively the dynamics of the swarm with respect to internal preference factors of higher priority, the performance of the DMVD and DMMD strategies is likely to deteriorate. For example, in the foraging scenario represented in Fig. 9.1, the site with the best construction materials is also the farthest from the retrieval area and its cost might affect the collective decision. A possible extension of the modular design methodology proposed in this book is represented by the design of modulation mechanisms that are able to compensate the negative effects of indirect modulation by augmenting the strength of the contribution of direct modulation as a function of environmental bias factors too. To achieve this

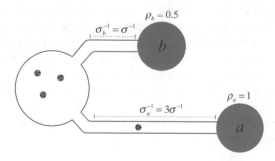

Fig. 9.1 Schematic illustration of a foraging scenario with two foraging sites (i.e., site a and site b). Site a contains construction materials that are twice as good as those contained in site b (i.e., internal preference) but is positioned three times farther than site b (i.e., negative environmental bias)

objective, agents would need to estimate the impact of environmental bias factors (e.g., the duration of the exploration state) and use this information when determining the contribution of direct modulation.

As a result of the proposed methodology, we were able to perform an extensive speed versus accuracy comparison of the DMVD and DMMD strategies by relying on predictive mathematical models. The result of this analyses is that the DMVD and DMMD strategies provide different design compromises. Their performance could be projected on a Pareto chart and would form a Pareto frontier with two points. A natural extension of our methodology that improves the choices of the designer is represented by the design of novel mechanisms for individual agents' decisions. For example, one could envisage a *Direct Modulation of k-unanimity Decisions* strategy (DMKD) to extend the Pareto frontier. A rather different and possibly more innovative approach is represented instead by the idea of behavioral heterogeneity. That is, to design swarms composed of different sub-populations of agents each executing a different collective decision-making strategy. This approach provides a much higher granularity of design solutions and would allow the designer to choose from a more complete Pareto frontier. The reader might note that the same objective can also be achieved by letting agents choose at random which individual decision-making mechanism to apply.

In Chap. 3, we identified four fundamental mechanisms that are necessary for consensus achievement. However, this set of mechanisms is far from being complete. For example, a factor that has a prominent role in shaping the collective dynamics is spatiality because it constrains the interactions among agents and between agents and their environment. The execution of control rules by individual agents in their environment creates an interaction network with volatile structure. In turn, the topology of this network influences the collective decision-making process by shaping the flow of information within the swarm. From this perspective, it is natural to think about the possibility of finding additional fundamental mechanisms that could extend the proposed design methodology. For example, the performance of a strategy could be improved by considering mechanisms that promote the formation of certain interaction structures able to maximize the flow of information within the swarm. To this end, network science and information theory offer tools to support the development of future research. The former provides theoretical frameworks to model the dynamics of the swarm at the level of the agents' interaction network, e.g., adaptive network models (Couzin et al. 2011; Huepe et al. 2011; Zschaler et al. 2012). The latter gives access to the means necessary for measuring the flow of information in the network; these means include measures of information transfer that give both a local perspective (Lizier et al. 2008), i.e., between pairs of nodes of the network, and a global perspective (Balduzzi and Tononi 2008), i.e., between pairs of macroscopic states of the overall network.

References

D. Balduzzi, G. Tononi, Integrated information in discrete dynamical systems: motivation and theoretical framework. PLoS Comput. Biol. **4**(6), 1–18 (2008)

I.D. Couzin, C.C. Ioannou, G. Demirel, T. Gross, C.J. Torney, A. Hartnett, L. Conradt, S.A. Levin, N.E. Leonard, Uninformed individuals promote democratic consensus in animal groups. Science **334**(6062), 1578–1580 (2011)

C. Huepe, G. Zschaler, A.-L. Do, T. Gross, Adaptive-network models of swarm dynamics. New J. Phys. **13**(7), 073022 (2011)

J.T. Lizier, M. Prokopenko, A.Y. Zomaya, Local information transfer as a spatiotemporal filter for complex systems. Phys. Rev. E **77**, 026110 (2008)

M.A. Montes de Oca, E. Ferrante, A. Scheidler, C. Pinciroli, M. Birattari, M. Dorigo, Majority-rule opinion dynamics with differential latency: a mechanism for self-organized collective decision-making. Swarm Intell. **5**, 305–327 (2011)

G. Zschaler, G.A. Böhme, M. Seißinger, C. Huepe, T. Gross, Early fragmentation in the adaptive voter model on directed networks. Phys. Rev. E **85**, 046107 (2012)

Appendix A
Background on Markov Chains

We review the concepts underlying the theory of time-homogeneous Markov chains that are relevant for the understanding of this book. The discussion presented in this appendix focuses on absorbing time-homogeneous Markov chains over a finite state-space. We briefly introduce the formalism of Markov chains, discuss the memory-less property underlying this formalism, and illustrate how states of a chain can be organized in different equivalence classes. We focus on absorbing Markov chains and illustrate how to derive the canonical form of the chain as well as a number of statics of interest in this book.

A.1   Time-Homogeneous Markov Chains

In the following, we summarize the fundamental principles of Markov chain; for a thorough introduction to Markov Chains, we refer the reader to "Finite Markov Chains" by (J. G. Kemeny and J. L. Snell 1976) and "Markov Chains" by (J. R. Norris 1997). Let $\mathbb{N}$ represent the set of naturals. Each element $i \in \Omega$ is called a *state* and the finite set $\Omega \subset \mathbb{N}$ is called the *state-space*. We say that the vector $\lambda = (\lambda_i : i \in \Omega)$ is a probability distribution on Ω if $0 \leqslant \lambda_i \leqslant 1$ for all $i \in \Omega$ and $\sum_{i \in \Omega} \lambda_i = 1$. We consider the *random variable* X which takes values in Ω and let

$$\lambda_i = P(X = i), \tag{A.1}$$

where the function P gives the probability of the event $X = i$. The random variable X models a random state which takes value i with probability λ_i and λ defines the probability distribution of X.

Let $P = (p_{ij} : i, j \in \Omega)$ represent a stochastic matrix, that is, a matrix in which every row $(p_{ij} : j \in \Omega)$ is a probability distribution. We say that the sequence of random variables $\{X_\vartheta : \vartheta \in \mathbb{N}\}$ is a *Markov chain* with *initial distribution* λ and *stochastic transition matrix* P if the following two conditions are satisfied:

© Springer International Publishing AG 2017
G. Valentini, *Achieving Consensus in Robot Swarms*, Studies in Computational
Intelligence 706, DOI 10.1007/978-3-319-53609-5

i. X_0 has probability distribution λ,
ii. for each time-step $\vartheta \in \mathbb{N}$ conditioned to the fact that $X_\vartheta = i$, we have that $X_{\vartheta+1}$
   has probability distribution $(p_{ij} : j \in \Omega)$ and is independent of $X_0, \ldots, X_{\vartheta-1}$.

The first condition states that the probability distribution λ defines the distribution
of the initial state X_0 of the chain at step $\vartheta = 0$. The second condition defines the
Markov property of a stochastic process—the lack of memory. A stochastic process
has the Markov property if, at any time-step, the future state of the process depends
only on the current state and is independent of its past. Given $X_\vartheta = i$, the entry p_{ij}
of the stochastic transition matrix P gives the probability that the next state $X_{\vartheta+1}$
will correspond to j. Note that matrix P is independent of the time-step ϑ. This time
independence is referred to as the *time-homogeneous property* of the Markov chain.
In general, the probability to find the stochastic process in state $j \in \Omega$ at time-step ϑ
will be equal to

$$P(X_\vartheta = j) = (\lambda P^\vartheta)j, \qquad (A.2)$$

where the function $(\cdot)j$ returns the j-th element of a vector.

The state-space Ω of the Markov chain can be divided into *equivalence classes*
of states—also known as *communicating classes*. Within an equivalence class, the
stochastic process can transit from any initial state to any other state, although not
necessarily in one time-step. If, from a state i of the equivalence class $\mathscr{A}$ it is not
possible to go to a state j of any other equivalence class $\mathscr{B}$, then $\mathscr{A}$ is said to
be *closed* and their states are called *ergodic states*. In the opposite case, we say
that the states are *transient states*. As a consequence of the previous condition, we
have that (1) if a stochastic process leaves a certain transient class it will never return
to that class and (2) if the stochastic process enters a closed class, it will never
leave that closed class. In particular, an ergodic class that consists of a single state i
corresponds to a state that cannot be left by the stochastic process; this state is called
absorbing state. An absorbing state i is characterized by probability $p_{ii} = 1$ and
probabilities $p_{ij} = 0, \forall j \neq i$. A Markov chain that has at least one absorbing state is
called absorbing Markov chain. The stochastic process moving along an absorbing
Markov chain will eventually enter (or be absorbed in) one of the absorbing states.

A.2 Analysis of Absorbing Markov Chains

We consider an absorbing, time-homogeneous Markov chain $\{X_\vartheta : \vartheta \in \mathbb{N}\}$ with
state-space $\Omega = \{1, \ldots, w\}$, $w \in \mathbb{N}$. The state-space Ω includes a number $r \geqslant 1$ of
absorbing states. We remark to the reader that, during the development of this book,
we make use of absorbing states to identify macroscopic configurations of the agents
of the swarm that are of interest to the designer, e.g., the achievement of consensus
on a particular opinion. As a consequence, absorption represents the completion of
the swarm's task.

Once the Markov chain is completely defined, the stochastic transition matrix P can be used to answer a number of questions regarding the performance of the system it models. To this end, the first step of the analysis consists of finding the *canonical form* of matrix P. This objective can be achieved by reordering the states in Ω so that to write the stochastic transition matrix P as

$$P = \left(\begin{array}{c|c} I & O \\ \hline R & Q \end{array} \right). \tag{A.3}$$

In Eq. (A.3), Q is a $(w-r) \times (w-r)$ matrix that contains the transition probabilities between transient states; R is a $(w-r) \times r$ matrix that contains the transition probabilities from a transient state to an absorbing state; matrix O consists entirely of 0's; and matrix I is the identity matrix with size $r \times r$ which identifies the absorbing states of the Markov chain. Note that, as a consequence of the absorbing nature of the Markov chain, the entries of Q^ϑ converge to 0 as the time-step $\vartheta \to \infty$. That is, the probability to find the process in a transient state vanishes as the time passes. This result provides sufficient conditions for the existence of the inverse of matrix $I - Q$, which is called *fundamental matrix* and is given by

$$F = (I - Q)^{-1} = I + Q + Q^2 + \cdots = \sum_{k=0}^{\infty} Q^k. \tag{A.4}$$

Each entry f_{ij} of the fundamental matrix F provides to the mean number of time-steps that a process started in the transient state i spends in the transient state sj.

By means of the canonical decomposition of P we can derive a number of interesting macroscopic quantities regarding the dynamics of a swarm. The first of these quantities is the probability that a system initially started in state $X_0 = i$ will eventually enter the absorbing state $X_\vartheta = j$ at a certain time-step ϑ. If we consider all possible initial states i and all possible absorbing states j, we have that the set of absorption probabilities is determined by the matrix

$$B = FR. \tag{A.5}$$

Each entry b_{ij} of B gives the absorption probability for the pair (i, j) of initial and absorbing states.

In addition to the absorption probabilities, we can study the time to absorption, that is, the number of time-steps necessary to a stochastic process moving along the chain to enter an absorbing state. Let us denote with τ the random variable that counts the number of time-steps before absorption. The expected value $\hat{\tau}$ and the variance $\hat{\tau}_2$ of τ are given by equations

$$\hat{\tau} = F\boldsymbol{\xi}, \tag{A.6}$$

and,

$$\hat{\tau}_2 = (2\mathbf{F} - \mathbf{I})\hat{\tau} - \hat{\tau}_{sq}. \tag{A.7}$$

In the above equations, the symbol $\boldsymbol{\xi}$ identifies a column vector of all 1's and the vector $\hat{\tau}_{sq}$ corresponds to $\hat{\tau}$ with squared entries. The i-th entry of the vectors $\hat{\tau}$ and $\hat{\tau}_2$ gives, respectively, the expectation and the variance of the time necessary before absorption for a system initially started in state i.

Finally, we can derive the cumulative distribution function $P(\tau \leqslant \vartheta; i)$ of the time to absorption (and consequently, the probability mass function $P(\tau = \vartheta; i)$) for any initial state $X_0 = i$. The cumulative distribution function is obtained as the infinite series

$$P(\tau \leqslant \vartheta; i) = 1 - \sum_{j \in \Omega} \mathbf{Q}_{ij}^{\vartheta}, \text{ for } \vartheta \to \infty. \tag{A.8}$$

The term $\sum_{j \in \Omega} \mathbf{Q}_{ij}^{\vartheta}$ in Eq. (A.8) gives the probability that the stochastic process will be in a transient state at step ϑ. The complement of this value provides the probability that the process has entered an absorbing state before time-step ϑ.

References

J.G. Kemeny, J.L. Snell, *Finite Markov Chains* (Springer, New York, 1976)
J.R. Norris *Markov Chains* (Cambridge University Press, New York, 1997)

Printed in the United States
By Bookmasters